Basi
structural
behaviour

understanding structures from models

Barry Hilson

BSc(Eng), PhD, CEng, FICE, FIStructE, FIWSc
Professor of Civil Engineering, University of Brighton

Thomas Telford, London

Published by Thomas Telford Services Ltd, Thomas Telford House, 1 Heron Quay, London E14 4JD.

First published 1993.

A catalogue record for this book is available from the British Library.

ISBN: 0 7277 1907 6

Typeset in Great Britain by MHL Typesetting Ltd.

Printed and bound in Great Britain by The Eastern Press Ltd

Acknowledgements

The Author wishes to thank the following for permission to include copyright photographs.

The Council of the Institution of Civil Engineers for Fig. 1.16 which appeared in the paper 'Experimental verification of the strength of plate girders designed in accordance with the revised British Standard 153: tests on full-size and on model plate girders', by E. Longbottom and J. Heyman, *Proc. Instn Civ. Engrs,* Part III, 1956, **5**, Aug.

The Director of the Building Research Establishment for Figs 1.14 and 3.12.

British Steel for Figs 2.19, 2.20, 2.21, 4.22, 5.18, 5.19, 6.10, 6.11, 7.13 and 8.4.

Rainham Timber Engineering Co. Ltd for Figs 4.21 and 7.14.

K. Thomsen Esq., International Steel Consulting Ltd, Denmark, for Fig. 2.18.

The Timber Research and Development Association for Figs 4.18 and 8.5.

Brecht-Einzig Ltd for Fig. 6.9.

The Cement and Concrete Association for Fig. 7.12.

The British Constructional Steelwork Association Ltd for Figs 3.13 and 4.19.

Wolf Systems Ltd for Fig. 5.17.

Contents

Introduction   **1**

1 Basic definitions   **5**
Introduction, Force, Reaction, Tension, Compression, Moment of a force, Couples, Equilibrium, Shear, Torsion, Bending moment and shearing force, Stress, Principal stress, Young's modulus, Second moment of area, Buckling, Local buckling, Lateral torsional buckling.

2 General stability   **18**
Introduction, Thin wall elements, Counterbraced panels, Other forms of shear wall, Multi-storey structures, Multi-bay structures, Practical examples, Exercises.

3 Column elements   **31**
Introduction, Slender columns, Effective lengths, Thin-walled columns, Battened columns, Practical examples, Exercises.

4 Beam elements   **46**
Introduction, I-beams, Lattice girders, Practical examples, Exercises.

5 Arches and roof trusses   **63**
Introduction, Arches, Roof structures, Practical examples, Exercises.

6 Grids 78

Introduction, Stiffness, Square grids, Diagrids, Practical examples, Exercises.

7 Folded plate structures 90

Introduction, Simple model forms, Practical examples, Exercises.

8 Composite behaviour 100

Introduction, Beam and slab systems, Exercises.

9 Miniature bridge design exercise 107

Introduction, Miniature lifting bridge: brief and specification, Comments

Index 111

Introduction

The structural design process may be represented simplistically by the accompanying flow chart.

The first stages are concerned with identifying possible structural solutions to meet the client's requirements. The choice of the best form is then made, and this is closely related to the structural material to be used. For example, if a truss form is to be adopted it could be constructed from steel, timber, alloys or even concrete, but it would not be possible in brickwork. Rules of thumb can be employed to identify approximate sizes for the individual members of the structure.

Having chosen a form, the loads that the structure has to carry may be assessed. These may be superimposed loads due to such things as snow, people, equipment, wind action, earthquake, etc., but it must never be forgotten that a structure must also support its own weight as well as superimposed loads; this may frequently be a high proportion of the total. Self-weight loads are often called 'dead loads' since they remain in one position throughout the life of the structure. Superimposed loads are then referred to as 'live loads'.

A preliminary analysis may then be conducted to estimate the compression and tension forces, bending moments, shearing forces and torsions that will be generated in the parts of the structure. A check must be made to ensure that the member sizes are sufficient to resist these effects. They can be modified if necessary and a fresh analysis and a fresh check made. This process ensures that the structure has sufficient strength to resist the applied loads.

A clear picture of the structure will by now have been developed, and the next stage will be to assess how the structure will behave under load. Not only must a structure be strong enough to resist the applied loads

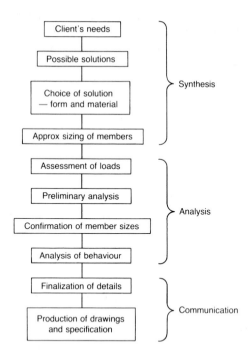

without failure, but it must also perform in a manner that is acceptable to the users. This is the 'serviceability' part of the design. These studies must show the acceptability of the levels of deflection, sway, vibration and cracking, for example.

Finally, the working drawings and specification must be produced, and these form the main communication links between the designer and the contractor.

In practice, the design process is not the simple linear process described above; some iteration is bound to occur, but the principles remain the same. Safety and economy are often conflicting requirements, and the balance between the two often represents the designer's biggest challenge.

If many thousands of a given design are required, then a range of possible designs could be produced and costed, and a design could be chosen which satisfies the client's requirements in terms of safety, economy, serviceability, aesthetics, etc. The development costs would then be spread over the total number to be made and would not be high per unit. However, in structural engineering, most projects are 'one-off' structures and the costs of producing a range of alternative designs is prohibitive. Consequently, the designer must quickly choose an appropriate form and material, and this choice must be both economical and satisfactory. It will probably not be the optimum solution but it must be close to it.

It is of the utmost importance, therefore, that anyone concerned with the design of structures should have the ability to visualize how a structure will behave in a given set of circumstances, and how the form of the structure will influence this behaviour. The designer must develop an intuitive feeling for structural behaviour so that when the all important choice of the structural form to be used is made, it is made correctly. The satisfactory outcome of a project in terms of aesthetics, economy and safety depends upon this important decision. Too often, designers make this decision hastily, and it is not until much later, after a considerable amount of further work, that they realize that a different form would have produced a more satisfactory solution.

New materials demand new structural forms if they are to be used efficiently, such forms being dictated by the material properties. It is not sufficient to use forms that have been found satisfactory in the past for more traditional materials. New thinking and clear understanding is required. Pier Luigi Nervi in his book *Structures* said '. . . we must perfect and go beyond the scientific mathematical stage of our knowledge and reach a stage of intuitive knowledge'.

How then can the designer obtain this intuitive feeling for structure? It is certainly not an inbred quality, something that a person either has or has not; rather, is it something acquired through experience. As children, through a process of trial and error, we gradually learn which arrangements of piled-up wooden blocks will produce stable structures, so that we now know intuitively how to construct that sort of simple structure.

Experience, then, is the key to the correct choice of structural form. Experience comes in two main forms: (a) the collective experience of the profession as a whole which is disseminated through papers in the professional and technical journals; and (b) personal experience that the designer has acquired through his or her professional career. To develop personal experience the designer could apply trial and error processes which would increase intuitive understanding of how structures behave. Trial and error processes cannot be applied to actual structures for obvious reasons, and therefore the study of model structures comes to mind as a very rapid means of acquiring personal experience.

This book uses illustrations of model structures under load to assist in the development of an understanding of structural behaviour. The models are made from balsa wood, balsa cement, cotton, string and thin card. The reader could extend the studies by making and testing the models, or the book may just be used as an introductory text on structural behaviour. If models are built, since some of the studies involve a comparison between the behaviour of similar forms, it is advisable that effects due to variation in material properties should be eliminated. This may be achieved by stress grading the balsa wood sticks and using only sticks of a similar strength. Since the strength of timber is directly proportional to its specific gravity, stress grading may be carried out by weighing and grouping the sticks according to their weight per unit length.

The models could be made and tested by small groups of students, say up to four per group. In this way students progress more rapidly by learning from each others' ideas, and they also develop other skills related to communication, compromise and organization. It is considered that the studies set out in this book are suitable for students following GCSE and 'A' level courses in design and technology, BTEC courses, and degree courses in structural engineering, civil engineering, building and architecture.

1 Basic definitions

1.1 Introduction

This chapter presents simple definitions of terms that are commonly employed in structural engineering. The aim is to help the reader who is unfamiliar with this subject to understand the discussions that follow in the later chapters. Further definitions are included in the text as appropriate.

1.2 Force

A force may be defined as an action on a body, acting in a specific direction which tends to move the body or, if it is already moving, to change its rate of movement. A force is normally measured in Newtons (N) or kilo Newtons (kN).

   The weight of a body is one kind of force: a measure of the action of gravity on the quantity of matter (mass) of the body.

1.3 Reaction

Consider a mass with a weight W resting on a table, as shown in Fig. 1.1.

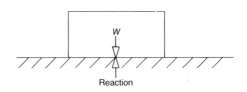

Fig. 1.1. *Weight force balanced by equal and opposite reaction force*

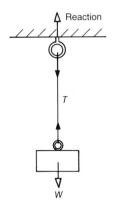

Fig. 1.2. *Weight force transmitted to reaction point (anchor) by tension*

The mass does not move under the attraction of gravity, and consequently the weight force must be cancelled by an equal and opposite force from the table, called the reaction force.

1.4 Tension

Figure 1.2 shows an object hanging from a piece of string. W is the force exerted by gravity on the mass of the object. Since the object does not move, the anchor must be applying an upward reaction force equal to W, so that the resultant force is zero.

By considering in turn the forces acting on the object and the anchor (Figs 1.3(a) and (b)) it may be seen that the string exerts an upward pull force T equal to W on the object and a downward pull T equal to W on the anchor. This force T is transmitted through the string, causing it to stretch. The string is said to be in a state of tension.

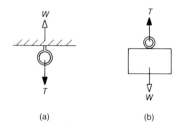

(a)   (b)

Fig. 1.3. *Equal and opposite forces acting on weight and on anchor*

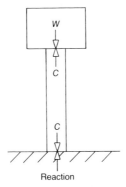

Fig. 1.4. *Weight force transmitted to reaction point by compression*

1.5   Compression

If an object with weight W is placed on top of a wooden prism resting on a table, as shown in Fig. 1.4, then the table must provide an upward reaction force, equal to W, in order to prevent movement.

At the top of the prism there must be an upward force C to balance W, and at the bottom a downward force C to balance the reaction. In each case this force C is equal to W and is transmitted through the prism. The prism is said to be in a state of compression.

1.6   Moment of a force

A plan view of a door is shown in Fig. 1.5. In order to open the door, a push force P is applied, causing the door to rotate about its hinge. The

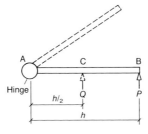

Fig. 1.5. *Turning moments applied to door*

turning effect due to a force is called the moment of a force and its magnitude depends on both the magnitude of the force and the distance of the force from the hinge or turning point. The moment of a force is measured by multiplying the magnitude of the force by the distance from the turning point, measured perpendicular to the line of action of the force.

In Fig. 1.5 the moment about A of the push force P applied at B is $P \times h$. The moment of a force is typically measured in Nmm. If a push force Q is applied at C, then the moment of the force about A would be $Q \times h/2$.

Since the turning effort required to open the door is the same for any position of the push force, it follows that $P \times h = Q \times h/2$.

$$\therefore Q = 2 \times P$$

Try this for yourself when opening a door by pushing it at a point well away from the hinges, and then at a point close to the hinges. A much greater force is required for the second procedure.

Another way in which the effect of the moment of a force may be felt is to hold a stick horizontally at one end and to hang a weight on the stick close to the hand. Now move the weight further away from the hand, keeping the stick horizontal. The moment of the force about the hand will increase as the distance increases, and the hand will need to apply an increasing resistance to the turning effect in order to keep the stick horizontal.

1.7   Couples

In Fig. 1.6 the turning moment of the force P about a hinge at A is equal to $P \times h$. If the whole body is not to move vertically there must be another force acting at A which is equal and opposite to P. Thus it may be seen

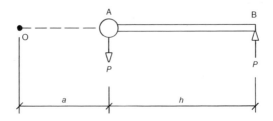

Fig. 1.6. *Turning couple*

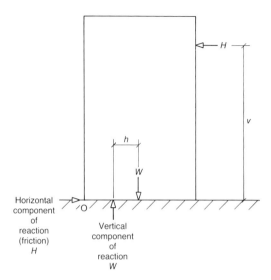

Fig. 1.7. *Equilibrium of block loaded with lateral push force*

that a turning effect comprises two equal forces spaced apart. This combination is called a couple, and its magnitude is measured by multiplying the force by the perpendicular distance between the two forces, equal to $P \times h$ in Fig. 1.6.

It may be noted that the magnitude of a couple is always the same, irrespective of the point about which moments of forces are being calculated. For example, in Fig. 1.6 the resultant moment of the forces about O is $P \times (a + h) - P \times a$ (negative because it is in the opposite direction), which still yields the value $P \times h$.

1.8   Equilibrium

A body is in a state of equilibrium when the effects of all the forces acting on it are balanced so that it does not move. Any downward forces must be balanced by upward forces, and any horizontal forces must also negate each other. There still remains the possibility, however, that the body could rotate; this is illustrated in Fig. 1.7.

The downward weight of the block is balanced by the upward reaction from the table. The horizontal force H is balanced by the frictional force developed between the table and the block. The overturning couple

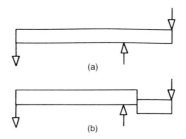

Fig. 1.8. *Example of shearing action; this tendency is present on all vertical sections of body*

developed by H and the frictional force (equal to $H \times v$) is balanced by the righting couple developed by W and the reaction force (equal to $W \times h$).

Note that the position of the reaction force varies with the magnitude of H in order that the necessary righting couple can develop. If $H = 0$ then the reaction force will align with W, since no righting couple would be required. As H is increased, the position of the reaction force will move towards O in order to increase the righting couple. Once it has reached O, the block will be on the point of overturning, and it will overturn if H is increased further since the position of the reaction force would now need to be outside the block where there is nothing for it to react against.

The above sequence of events will occur provided that the necessary frictional force, equal to H, can be developed between the block and the table. If not, the block will slide once H has reached the maximum value of the friction force that can be developed.

The requirements for the equilibrium of a body may be summarized as follows

(a)   the total upward force must equal the total downward force
(b)   the total horizontal force acting towards the left must equal that acting towards the right
(c)   the sum of the clockwise turning couples must equal the sum of those acting anticlockwise.

1.9   Shear

When a force system acts upon a body such as that shown in Fig. 1.8(a) then there is a tendency for parts of the body to slide relative to the remainder, as shown in Fig. 1.8(b). This sliding action is called shear.

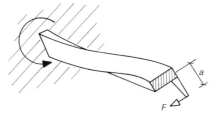

Fig. 1.9. *Torsion, or twisting action*

1.10 Torsion

When the forces acting on a body cause it to twist, then this is called torsion, and it is illustrated in Fig. 1.9. The amount of torsion is measured in the same way as the moment of a force, i.e. the magnitude of the force multiplied by the distance to the force measured perpendicular to its line of action. In the example shown this would be $F \times a$. For equilibrium, the torsional moment must be balanced by an equal and opposite moment supplied by the end anchorage in the example. The member is in a state of torsion, the torsional moment being transferred from one end to the other.

1.11 Bending moment and shearing force

Consider the member shown in Fig. 1.10(a), which carries two forces (or loads) F_1 and F_2 balanced by the two reaction forces R_1 and R_2. All

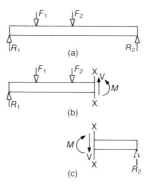

Fig. 1.10. *At each cross-section in member there will be internal moments and shearing forces to balance external moments and shearing forces acting to one side, or the other, of cross-section*

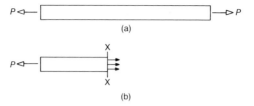

Fig. 1.11. *Tensile stress*

portions of the member are in equilibrium, so if part of the member is considered (Fig. 1.10(b)) then this part must also be in equilibrium and, therefore, there must be an internal moment at X—X to balance the moments of the external forces and an internal force to balance the external forces. These are shown as *M* and *V*, respectively.

The internal moment is called a moment of resistance, and it indicates the internal strength required from the member in bending to resist the resultant moment of the external forces which is called the bending moment. The internal force is called a shear resistance and shows the internal strength required from the member in shear to resist the resultant external force which is called the shear force.

For equilibrium, *M* must balance the total algebraic sum of the moments of the forces acting to the left of X—X, and *V* must balance the total algebraic sum of the forces acting to the left of X—X. The same results would have been obtained had the portion of the member to the right of X—X been considered, as shown in Fig. 1.10(c), except that the internal balancing force *V* would have acted in the opposite direction.

The bending moment at a section may therefore be defined as the algebraic sum of the moments of the forces acting to *one side* of the section under consideration. Similarly, the shearing force may be defined as the algebraic sum of the forces acting to *one side* of the section under consideration. Shearing action always comprises two equal and opposite forces.

1.12   Stress

Stress is a measure of the intensity of an internal force produced by the application of an external force, and it is normally measured in N/mm^2. Depending on the way in which the applied force is resisted, there may be a tensile stress, compressive stress, bending stress, shear stress, etc. The simplest case, that of pure tensile stress, is illustrated in Fig. 1.11.

Considering the equilibrium of the portion of the tension member to the left of section X—X, the external tensile force P must be balanced by the sum effect of the internal stresses acting on the cross-section at X—X. In this case it is assumed that the stress is uniform over the cross-section, so that

$$P = \text{stress} \times \text{area of cross-section}$$

$$\therefore \text{stress} = \frac{P}{\text{area}}$$

Now consider two blocks of steel, square in cross-section and standing on a table, as shown in Fig. 1.12. Each block has the same volume and hence the same weight W, so the reaction force R from the table will also be the same in each case, equal to the weight force. However, due to the different areas of contact, the stress at the base of each block will be very different, as shown.

1.13   Principal stress

Frequently, a point in a structure is acted on by a combination of different types of stress. For stresses acting in two dimensions the resultant forces

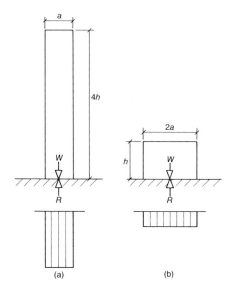

(a)   (b)

Fig. 1.12. *Effect of area upon intensity of stress: (a) stress* $= W/a^2$; *(b) stress* $= W/4a^2$

Fig. 1.13. *Crushing failure in short stocky compression member*

Fig. 1.14. *Overall buckling of bottom length of compression member in steel framework (Building Research Establishment: Crown copyright 1992)*

may be resolved into two principal directions, at 90° to each other, to produce two stresses which are purely tensile or compressive. These are called principal stresses and they represent the maximum and minimum tensile or compressive stresses at the point being considered. Similarly, for stresses acting in three dimensions there will be three principal stresses.

1.14   Young's modulus *E*

Young's modulus is the property of a material which measures its resistance to deformation in tension, compression or bending. This definition only applies if the material is stressed within its elastic limit, i.e. provided that the member regains its original shape once the forces producing the stresses and consequent deformation are removed. This is the situation that is normally assumed for structures under design loads.

Another name for Young's modulus is the modulus of elasticity. It is measured in N/mm^2.

1.15   Second moment of area *I*

The second moment of area is the geometric property of a cross-section which, when multiplied by Young's modulus for the material being used, provides a measure of a member's resistance to deflection under the action of bending. The product $E \times I$ is, consequently, referred to as the flexural rigidity of the member. I is measured in mm^4.

1.16   Buckling

When a short stocky member is loaded in compression it will usually eventually fail by crushing, as shown in Fig. 1.13. However, if the member is long and slender then, before the crushing capacity is reached, the member will bow sideways and will eventually fail in bending, as shown in Fig. 1.14. This latter action is known as overall buckling.

1.17   Local buckling

It sometimes happens that the member under compressive stress does not buckle overall (as shown in Fig. 1.14), but develops a localized deformation, as shown in Fig. 1.15. This is referred to as a local buckling failure, and it can frequently precipitate total collapse.

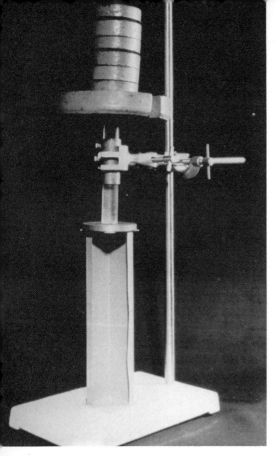

Fig. 1.15. *Local buckling on unsupported edges of model loaded in compression*

Fig. 1.16. *Lateral torsional buckling of steel girder*

1.18 Lateral torsional buckling

The upper part of the beam illustrated in Fig. 1.16 is in a state of compression, and therefore it tends to buckle like a column (Fig. 1.14). However, the lower part is in tension and remains straight, thus tending to restrain the column action to some degree. Eventually the compression zone may buckle sideways as shown, causing lateral bending and twisting of the beam. This is called lateral torsional buckling. Lateral buckling is accompanied by twisting of the section, and therefore more resistance is provided by sections that are stiff under torsion.

2 General stability

2.1 Introduction

For convenience, structures are normally divided into various individual elements and each element is designed to carry the loads imposed upon it. It is, however, extremely important that the designer should realize that these elements are frequently interdependent, and that there are occasions when the structure must be considered as a whole to appreciate the way in which it resists load.

For example, when wind forces act laterally on the side of a multi-storey building, these forces are transmitted by the cladding to the vertical framing elements, which in turn transmit them to the horizontal slab elements. These

Fig. 2.1. *Wall element offering negligible resistance to lateral force*

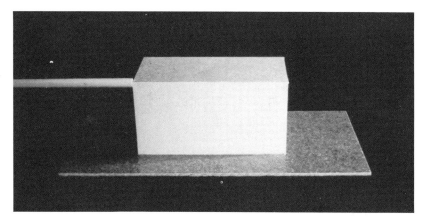

Fig. 2.2. *Force applied in plane of wall element is resisted strongly*

slab elements must therefore be supported against lateral movements, and such support may be provided by wall elements. The slab elements must be capable of transmitting the applied forces to the wall elements by virtue of their in-plane strengths, or alternatively bracing must be provided in the horizontal planes. Thus the forces are transferred to the walls, whence they are transmitted to the ground.

The stabilizing of the slab elements by the walls plays an important part in preventing lateral buckling of the beam elements and controlling the buckling length of column elements, as will be seen later. In general, support to slab elements is provided by a system of shear walls; the term 'shear wall' is used in this context in its widest sense to mean any two-dimensional structural system that is capable of transmitting in-plane shearing forces.

2.2   Thin wall elements

A thin wall element has negligible resistance to horizontal forces applied in the direction of its thickness, but considerable resistance to those applied in the direction of its length. This is illustrated in the models shown in Figs 2.1 and 2.2. Consequently, a single thin wall element in a structure does not provide general stability.

Providing two thin wall elements at right angles might at first sight appear to offer a general solution; each wall supplies resistance to the forces applied in one of the principal directions. However, if these forces are applied to the roof slab at the end remote from one of the walls, it will be noted that

Fig. 2.3. *Twisting occurs when force is applied away from junction of two walls*

torsional displacements occur, as shown in Fig. 2.3, and hence the structure will be relying for stability on the torsional stiffness of the L-shaped cantilever formed by the two walls. This is generally unreliable, especially with an open section such as an L-shape.

The provision of a third wall ensures that when a turning moment results from a load applied parallel to but away from the line of resistance of one of the shear walls, the two remaining walls provide a resisting couple, thus giving stability to the roof slab and eliminating torsional effects on the shear

Fig. 2.4. *Arrangement of three wall elements which resists forces applied to roof element in any direction*

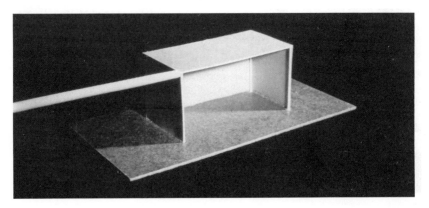

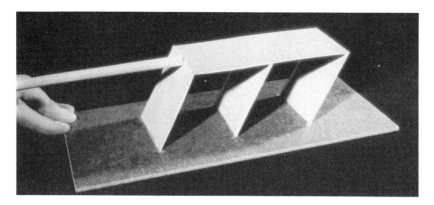

Fig. 2.5. *When all wall elements are parallel, system is ineffective*

wall assembly. This is illustrated in Fig. 2.4.

It may be seen, therefore, that a basic requirement for stability is the provision of *three* shear walls in each storey of a structure. There are certain constraints on the positioning of the walls, however. For example, the three shear walls must not all be parallel; this follows from the fact that a shear wall is only effectively able to resist load in its own plane, and if the walls were all parallel there would be no component available to resist a load applied perpendicular to the wall directions (Fig. 2.5). This would be true

Fig. 2.6. *When all planes of wall elements meet at single point, system is ineffective*

Fig. 2.7. *Single diagonal tie can stabilize this situation through its ability to transmit tension*

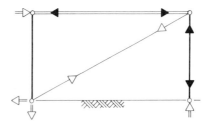

Fig. 2.8. *Forces involved in system illustrated in Fig. 2.7*

no matter how many parallel shear walls were provided.

Also, the lines of resistance, i.e. the tops of the planes containing the shear walls, must not meet at a point. If this were to be the case then any load applied other than through the intersection point would cause rotation of the roof slab, since none of the walls could provide a balancing moment about the intersection point. This type of behaviour is illustrated in Fig. 2.6. Even if the lines of resistance nearly intersect at a point, the resulting forces on the walls, to produce rotational equilibrium of the roof slab, will be considerable. A good solution is to have two of the shear walls parallel to one another, a reasonable distance apart, and the third wall perpendicular to the others.

The minimum requirements for the in-plane stability of the roof slab may be summarized, therefore, as three shear walls which are not all parallel, nor have lines of resistance which meet at a point. These minimum requirements are also applicable to any other element subject to multi-directional in-plane loading. In a horizontal beam element, for example,

Fig. 2.9. *Single diagonal tie in this situation cannot resist compressive force that is developed; it buckles and is ineffective*

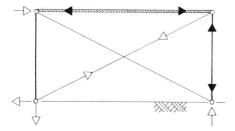

Fig. 2.10. *Use of two diagonal ties ensures that one is effective, irrespective of direction of applied in-plane force*

one end of the beam may be hinged, thus providing a vertical and a horizontal reaction component, and the other end supported on a roller or sliding support, thus providing a vertical reaction component.

It should be realized that if more shear walls are provided than the minimum required for structural stability, then an increase in structural stiffness will result, i.e. a greater resistance to lateral displacements. Shear walls should be located at strategic positions throughout a structure so that the forces can easily be transferred to them.

2.3 Counterbraced panels

By considering the plane frame shown in Fig. 2.7 it may be seen that under the action of the applied push the diagonal cotton member will be in a state of tension and this will transmit the applied force to the ground, i.e. the frame appears to be acting as a shear wall; this is shown diagrammatically in Fig. 2.8. If, however, the load is applied in the opposite

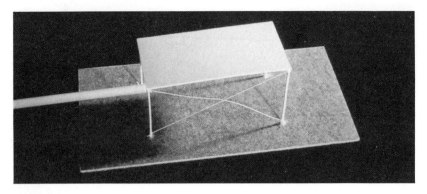

Fig. 2.11. *Action of counterbraced panel: note that one tie is doing all the work, the other having buckled and become ineffective; roles would be reversed if force were applied in opposite direction*

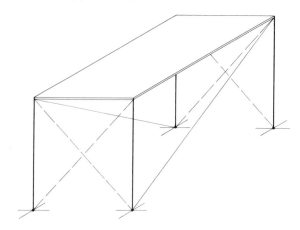

Fig. 2.12. *Divided counterbraced panel*

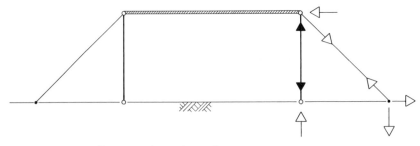

Fig. 2.13. *Externally counterbraced panel*

Fig. 2.14. *By fixing joints at top of columns, angular changes are resisted and shear transmitted by portal frame action*

direction, then the diagonal member will need to be able to resist a compressive force, which is not possible with a simple cotton tie, and collapse will result (Fig. 2.9). By adding another diagonal tie, as shown in Fig. 2.10, stability is once again restored. Such a system may be referred to as a counterbraced panel and, since it is capable of carrying in-plane shear forces in either direction, it forms an efficient shear wall.

Although the counterbraced panel appears to be statically indeterminate, in fact this is not the case, since no matter in which direction the force is applied one of the diagonal members will always buckle and thus effectively be eliminated from the structure, as may be seen in Fig. 2.11.

In bracing a structure, one of the counterbraced shear walls may be divided between two panels, as shown in Fig. 2.12, if this is more convenient. Alternatively, the diagonal ties may be positioned external to the panel being braced (see Fig. 2.13).

2.4   Other forms of shear wall

Other bracing arrangements capable of transmitting applied in-plane shearing forces, and hence of acting as shear walls, include rigidly jointed · frameworks such as portal frames (Fig. 2.14), rectangular frames braced with one diagonal capable of resisting both tensile and compressive forces (Fig. 2.15), vertical cantilevers and triangular frames.

Fig. 2.15. *Single diagonal can provide stability, provided it can resist both tension and compression forces*

2.5   Multi-storey structures

In a multi-storey structure the lateral wind loading will be divided by the external cladding between the various floor plates. In any given storey there must be a sufficient number of shear walls to transfer the cumulative shear force from the floor plate above to the floor plate below. Thus, each storey is similar to the simple structure examined above in which the shear was transferred from the upper level (the roof) to the lower level (the ground). Consequently, in a multi-storey structure three shear walls per storey will be required, and they will need to meet the positional constraints described above.

Fig. 2.16. *Dissimilar shear wall arrangements for model two-storey structure*

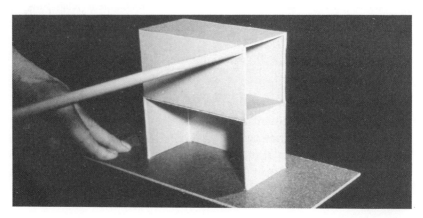

It is not necessary for the position of the three shear walls to be the same on plan in each storey. The transfer of shear through any one storey may be treated as an isolated problem. A dissimilar arrangement is shown in Fig. 2.16. However, it is frequently more convenient to adopt a repetitive arrangement, particularly when thin wall elements are being used, since the weight of the walls above any particular level may then be carried by the wall elements below. Moving down the building, the cumulative shear to be transferred through a storey does, of course, increase.

Lift shafts and other service shafts provide a very compact system of shear walls in a multi-storey structure. If, however, one is positioned towards one end of a building, another shear wall should be provided towards the other end to help to relieve the heavy torsional moments that would otherwise act on the shaft. An ideal solution for many high structures is to position the shaft centrally.

During construction, unless the actual framework of the building is to supply the shear wall requirement through the rigidity of its joints, it is essential that temporary bracing be provided until the permanent shear walls have been constructed and are able to resist load; otherwise, the wind forces on the unclad frame, and in particular the drag forces on any slab elements that may have been constructed, could be sufficient to cause a complete collapse of the structure. Temporary bracing normally takes the form of steel ropes arranged either to counterbrace panels, or to provide external diagonal ties, or both.

2.6 Multi-bay structures

In multi-bay single-storey structures, provided that the tops of each bay are interconnected, it is not necessary to provide shear walls in each. A

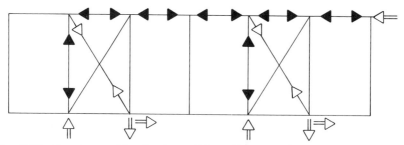

Fig. 2.17. *Concentrated bracing in multi-bay side walls*

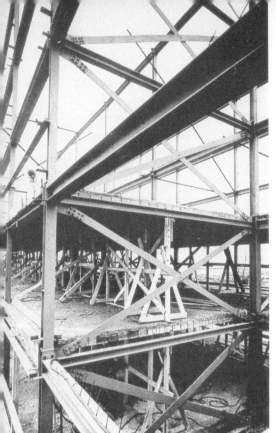

Fig. 2.18. *Counterbraced panel shear walls used for a steel-framed office building*

Fig. 2.19. *Unusual use of counterbracing to stabilize 338 m high building*

Fig. 2.20. *Steel-framed school building stabilized during erection with 12 mm diameter steel ropes; both counterbraced and external ties may be seen*

Fig. 2.21. *Hotel building with central, reinforced concrete, services shaft providing compact arrangement of shear walls*

typical arrangement is shown in Fig. 2.17 in which the penultimate bays of a series are counterbraced. The shear load to be carried will be transmitted by the connections and shared equally between the two braced panels. This form of bracing for a series of bays is preferable to an arrangement that braces every bay, not only because it is cheaper, but also because it produces a more flexible structure which is able to accommodate variations in temperature without the additional stresses that would occur in a more rigid structure.

2.7   Practical examples

Figures 2.18–2.21 illustrate applications of a variety of forms of shear wall.

2.8   Exercises

Construct a 150 mm × 75 mm rectangle by glueing together four lengths of 1·5 mm square balsa stick. Glue a sheet of thin card to this rectangle and a 75 mm length of 1·5 mm square balsa stick to each corner, thus forming a table framework. Lightly glue the bases of the four corner columns to a rigid board. The resulting model may then be considered as a single-storey single-bay structure with a horizontal roof slab and with the columns lightly jointed at each end.

Apply slight horizontal finger pressure to the roof level of the model and note the extreme flexibility of the structure.

Now fill one side at a time with thin card walls, glueing the card to the balsa framework and to the baseboard, until sufficient walls have been placed to ensure stability of the roof slab against any direction of horizontal force applied at roof level. Note the minimum number of walls that is necessary.

Build a new framework and repeat the exercise by now using only cotton ties placed in the planes of the side and end frames. Further studies may be made by allowing the ties to be placed in any position, either within the space of the structure or external to the structure.

3 Column elements

3.1 Introduction

A column is essentially a vertical member designed to transmit a compressive load. Being a compression member, it is reasonable to suppose that a column would fail by crushing of the material when the load reached a high enough value, but for most columns failure occurs at a lower load than the crushing strength; this is because most columns are relatively slender, i.e. they are long in relation to their lateral dimensions. It is generally observed that when a slender member is loaded in compression, as for example when a slender stick of timber is leaned on rather heavily, it will bow sideways or buckle, and if the load is then increased further the member will eventually fail in bending.

If, on the other hand, a stocky column is used, one with a low length-to-breadth ratio, then a crushing mode of failure is more likely than a buckling mode. For example, if a block of timber 70 mm × 70 mm × 140 mm high were loaded in compression, one could not imagine it failing by buckling (see Fig. 1.13).

Thus the normal intuitive feeling for common forms of compression elements says that length and lateral dimension play a part in determining the mode of failure that will result. Also, for a given section, there will be a critical length of compression member below which the member will be crushed and above which it will buckle.

3.2 Slender columns

A slender column made from 3 mm square balsa stick is shown, near to failure, in Fig. 3.1. The ends of the stick have been sharpened to a point

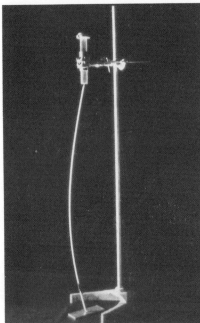

Fig. 3.1. *Pin-ended model column near to overall buckling failure*

Fig. 3.2. *Overall buckling mode for model column with top end pinned and bottom end fixed*

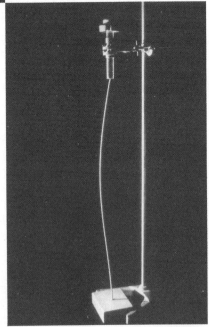

so that each is free to rotate, yet is prevented from moving laterally by friction. Thus, the ends are held in position but not in direction; this type of end condition is commonly called a hinged end or a pinned end.

If a number of such columns are tested, with a range of lengths from, say, 150 to 450 mm, then a study may be made of the effect of length on column strength. The study can be extended to include the effect of end conditions if it is repeated, with this time one, or both, ends of the columns prevented from rotating by glueing them into a block of wood, as shown in Fig. 3.2. The bottom end is held in position and in direction, and this end condition is commonly called encastré or a fixed end.

Typical results of such studies are shown in Fig. 3.3. In such studies it is necessary to use balsa sticks that are matched for strength. This may be achieved by weighing and using only sticks with a similar weight per unit length. From these it may be easily concluded that as the length of a column is reduced, the load it can carry is increased. It may also be concluded that preventing the ends of a column from rotating also increases the load-carrying capacity.

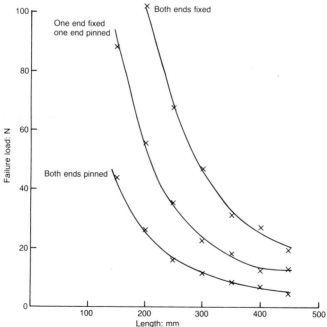

Fig. 3.3. *Buckling load–length relationships for columns with differing end conditions*

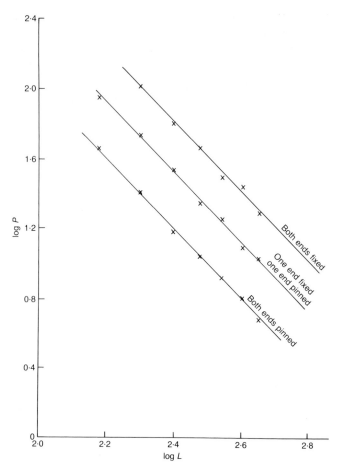

Fig. 3.4. *Logarithmic plot of results in Fig. 3.3 showing parallel relationships and hence same power index*

If it is assumed that a power, or exponential, relationship exists between the failure load and length for a given type of column, i.e. that $P \propto L^n$ then $P = $ constant $\times L^n$, and taking logarithms

$$\log P = \log(\text{constant}) + n \log L$$

This equation is of the form $Y = C + mX$, i.e. the equation of a straight line. By plotting $\log P$ against $\log L$ for the previous studies the value of the power index n may be found for each by measuring the slope of the resulting straight lines. This is shown in Fig. 3.4; in each case the slope of the straight lines is -2.

It may be concluded, therefore, that the buckling load is inversely proportional to the square of the length of the column and that this relationship is independent of the type of end conditions. The constant of proportionality will, however, change with the end conditions, being highest for the column with both ends fixed and lowest for the column with both ends pinned.

The above study used a given material and a fixed cross-section. If it were extended to other materials and other cross-sections, it would be found that the buckling load is proportional to the Young's modulus E for the material and to the second moment of area I for the cross-section. The full buckling equation then becomes

$$\text{buckling load} = \text{constant} \times EI/L^2$$

where the constant varies with the type of end condition.

3.3   Effective lengths

From the study in the previous section it is possible to determine the lengths of the various types of column that have the same failure load. These may then be expressed as ratios. For example, from Fig. 3.3 for a failure load of 40 N, the lengths are as follows: both ends fixed, $L_1 = 318$ mm; one end fixed, one end pinned, $L_2 = 228$ mm; both ends pinned, $L_3 = 158$ mm. Therefore

$$\frac{L_3}{L_2} = \frac{158}{228} = 0 \cdot 7$$

$$\frac{L_3}{L_1} = \frac{158}{318} = 0 \cdot 5$$

In other words, a column with both ends fixed and with length L buckles at the same load as a pin-ended column with length of $0 \cdot 5 \times L$. Thus any column may be reduced to an equivalent pin-ended one by multiplying its actual length by a factor which depends upon the degree of end restraint. These factors vary, in theory, from $0 \cdot 5$ for a column with both ends fixed to $2 \cdot 0$ for a column with one end fixed and the other totally free, i.e. neither fixed in position nor in direction like a flag pole. The equivalent pin-jointed length is normally referred to as the effective length, and the factor as the effective length factor.

The concept of an effective length is very useful in practice, since it enables all columns to be reduced to a common condition for comparison purposes. It also means that the only design chart (or table) that is required is that applicable to pin-ended columns. The problem that arises in practice, however, is that a perfectly pinned end or a perfectly fixed end, are extremely difficult to obtain. Constraints against rotation will vary between pinned and fully fixed so the question arises as to which factor should be applied to the actual length in order to obtain the effective length. There is no absolute answer to this question, and engineering judgement has to be employed. There is general agreement, however, that the lowest factor that should be used is $0 \cdot 7$.

As an example, consider a multi-storey framed structure. If the structure is to contain shear walls, rather than rely upon the rigidity of the joints for overall stability, then the floor plates will be held in space and they in turn will hold the columns in position at each floor level. Therefore, each storey height of column is at least as good as a pin-ended column, and so the factor to be applied to the storey height will not exceed $1 \cdot 0$. The question that remains concerns what degree of rotational constraint exists at the top and bottom of each storey height of column. Rotational constraint will be provided by the resistance offered to rotation by the continuing shaft of the column and by the connected beams at the level under consideration. If the beams at each end are relatively short and deep then they will offer considerable resistance to rotation, and the lowest practical factor of $0 \cdot 7$ would be applicable. If, on the other hand, the beams were relatively long and shallow then a higher factor, between $0 \cdot 7$ and $1 \cdot 0$, would be appropriate. It is usual in these cases to use a factor of $0 \cdot 85$.

3.4 Thin-walled columns

In columns made from metals it is normally more economical to form the cross-section from sets of interconnected plate elements rather than to use solid rectangular or circular shapes. A comparison of the behaviours and load-carrying capacities of these types of column may be made by taking a series of similar sheets of thin card, folding them into different shapes of cross-section and then loading them as columns. When closed sections are used the matching edges may be joined with adhesive tape.

In Fig. 3.5 the card has been left in its original thin rectangular form and has failed by overall buckling at a very low load. In Fig. 3.6 the card has

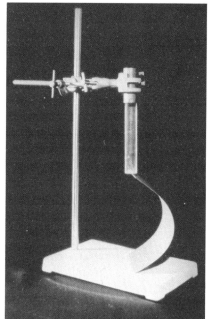

Fig. 3.5. *Negligible buckling capacity of thin rectangular form*

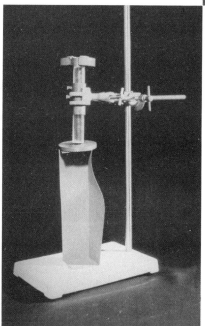

Fig. 3.6. *Local buckling in equal-angle section*

been shaped into the form of an equal angle. This column is stronger than the previous one and the mode of failure has changed. Local free edge buckling was followed by overall torsional buckling.

Other shapes that were investigated are illustrated in order of increasing load capacity. A channel section was used for the column shown in Fig. 3.7 and this failed in a similar manner to the angle section but at a higher load, local free edge buckling being followed by overall torsional buckling. The first closed section, illustrated in Fig. 3.8, was an equilateral triangle, and this failed by local ripple buckling of one side leading to overall lateral buckling. The square cross-section (Fig. 3.9) failed in a similar manner to the equilateral triangle but at a much higher load.

The strongest shape was the circular cross-section, shown in Fig. 3.10; this showed a massive increase in load capacity when compared with the original rectangular form, and failed by longitudinal concertina buckling.

There are many other shapes that could be investigated, and by listing

Fig. 3.7. *Local buckling in channel-shaped section*

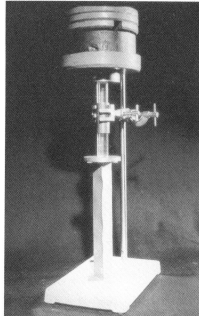

Fig. 3.8. *Closed triangular-section column showing longitudinal buckling wave pattern*

Fig. 3.9. *Longitudinal buckling wave pattern in closed square-section column*

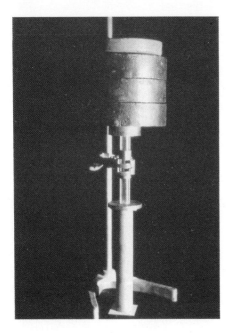

Fig. 3.10. *Large increase in load carried by circular-section column*

all shapes in order of increasing load-carrying capacity, a number of general observations may be made, as follows.

(a) Closed sections are more efficient than open sections, i.e. those with free edges.

(b) Sections with a greater number of plates are stronger. Since the perimeter of the cross-section is constant in all the columns, a greater number of plates means a smaller individual plate breadth. The circular section represents the ultimate limit of this series. The wavelength of the buckle wave can be shown to be a function of the plate breadth-to-thickness ratio. As the ratio decreases the wavelength also decreases. This is similar to reducing the buckling length of a column, as described above, and hence results in a higher load-carrying capacity.

(c) Local buckling can frequently develop in thin plate elements, and this can precipitate overall failure.

All the columns illustrated are thin-walled in relation to their overall dimensions, so that some have a high plate width to thickness ratio and, consequently, tend to fail by local buckling, and others (the open sections) have a low torsional rigidity and tend, therefore, to fail by overall torsional buckling. An increase in wall thickness could eliminate these two failure modes and enable higher loads to be carried. Failure would then be due to either overall buckling or crushing.

The above study clearly demonstrates the advantage of closed sections over open sections for columns, and it shows the circular cross-section to be the most efficient. However, if one considers the section which is most commonly employed for multi-storey steel-framed buildings it must be the Universal Column *I*-section, which is an open section! Why then do we use this less efficient section in preference to the highly efficient circular section? The circular section is more expensive but not excessively so. Some columns have to resist bending moments as well as carry compressive loads, and the I-section is more efficient as a beam (as will be seen later), but again this is not enough to counteract the considerable efficiency of the circular section. The main reason is that it is much easier to make connections to an I-section than to a circular section, for example for the attachment of the beams. Ease of construction is a factor that must never be overlooked at the design stage and can, in many cases, override other considerations such as design based on the least weight of constructional materials.

The circular section is, however, finding more and more applications as a compression member in cases where jointing techniques allow. For example, in roof truss construction the use of automatically cut and welded joints has enabled the highly efficient circular section to be employed for both the compression and tension members. No longer is it necessary to use the old, inefficient, angle iron sections joined via gusset plates and bolts or rivets.

3.5 Battened columns

A useful compromise for columns which retains the ease of connection of an open section and yet has some of the advantages of a closed section is to use what is called a battened column. This type of column employs two or four independent sections which are forced to work together as a single column by the attachment of batten plates at regular intervals along the length of the column. Fig. 3.11 illustrates a model column in which the four corner sticks of balsa wood have been connected by a series of card batten plates. This type of column is basically the same as the square section column in Fig. 3.9, but with the material concentrated at the corners

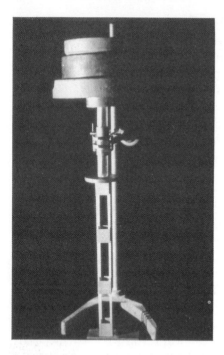

Fig. 3.11. *Model battened column*

so that buckling of the thin sheets is eliminated. However, in order that the four corner elements work together as one large member, they must be interconnected. Without the card batten plates each corner element would act independently and failure would occur by single curvature buckling at low load.

In order that batten plates may enable the corner elements to act compositely as one large member, they must be rigidly attached to the corner elements, and both the connection and plate must be capable of transmitting bending moments and shearing forces. The spacing must be such that individual buckling of the corner elements is eliminated.

Battened columns are normally employed when the unsupported buckling length of a column is high, and hence a large second moment of area is required to combat the long length. Fabrication costs can be high for this form of column.

Fig. 3.12. *Buckling of intermediate length column in steel framework (Building Research Establishment: Crown copyright 1992)*

Fig. 3.13. *Battened members used in construction of dock; in main columns two channel sections have been connected with batten plates*

Triangular lattices may be used as an alternative to batten plates as a means of achieving full composite action between corner elements. These were particularly popular in Victorian times, as may be seen by examining the structures in many of our larger railway stations.

3.6 Practical examples

Figures 3.12 and 3.13 illustrate some practical columns.

3.7 Exercises

From matched sticks of 3·0 mm square balsa wood take several pieces varying in length from 150 to 450 mm and taper each of the ends to a sharp point. Load each piece in compression as a column, using an arrangement such as that shown in Fig. 3.1, until failure occurs.

Plot a graph of failure load against 1/(length)^2.

Repeat the experiment but this time restrain the lower ends of the balsa wood columns against rotation, for example by glueing them into a block of wood.

Again plot failure load against 1/(length)^2.

What shape of graph should be obtained in each case?

The results will probably be slightly scattered, so make a list of all the small errors in the experiments which contribute to this scatter.

Take a sheet of cartridge paper 100 mm × 200 mm and apply a load parallel to the 200 mm side. Note the very low load at which overall buckling occurs.

Now take a series of similar sheets and fold each into a different shape of cross-section. Where closed sections are used join the edges with adhesive tape. To help maintain the shape of the resulting 200 mm high columns glue the top and bottom ends to 50 mm × 50 mm card plates.

Stand each column vertically and load to failure using a system similar to that for the previous exercise.

Note the mode of failure for each column and list the shapes in order of increasing load capacity.

4 Beam elements

4.1 Introduction

A beam is an element which carries load between supports by virtue of its resistance to bending and shearing action.

If two parallel vertical lines are drawn on the side of a block of rubber, and the rubber is then bent as shown in Fig. 4.1(b), it will be observed that the distance between the lines remains constant at the centre of the block, progressively increases towards the bottom and progressively reduces towards the top; i.e. the upper half of the block is in compression and the lower half in tension. The plane of the block which remains unstressed is called the neutral plane and it coincides with the centroid axis of the cross-section — mid-depth for a rectangular cross-section.

Therefore, for any element under bending, since the maximum compression and tension effects occur at the top and bottom of the section, it is logical to concentrate more of the material at these extreme surfaces. Any material in between will be less highly stressed and consequently used less efficiently.

Taking this idea to its limit, if all of the material could be concentrated at the two extreme surfaces, so that one area provided a compression resistance and the other a tension resistance, then all the material could be worked at its maximum capacity. For equilibrium, the tension and compression forces would be equal in magnitude and would form an internal couple to balance the applied bending moment. The arm of the couple would be the distance between the two forces, and the further the two areas of material, or flanges as they are called, were placed apart, the larger the bending moment that could be resisted.

This ideal state of affairs cannot be achieved in practice. In the first place,

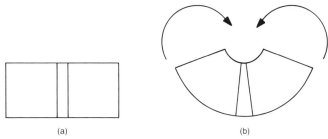

Fig. 4.1. *Rubber block: (a) before bending; (b) after bending*

the compression flange would need to be stabilized against the buckling modes of failure discussed in chapter 3; and secondly, in order that the two flanges may work together in resisting the applied bending moment, some interconnecting element must be provided to carry the vertical shearing forces due to the applied loads, and the corresponding horizontal shearing forces.

The presence of vertical shear becomes evident if one imagines a beam to be sliced vertically so that it resembles a row of books. Such a beam would be unable to carry vertical loads until some resistance to vertical shear was provided, for example, by glueing the slices together again.

In a similar way the presence of horizontal shear becomes evident if one imagines the beam to be sliced horizontally. If the resulting assembly were supported at each end and loaded centrally, the slices would slide over one another, each acting independently, as shown in Fig. 4.2. In an actual beam vertical and horizontal shearing tendencies will, therefore, be present and must be suitably resisted.

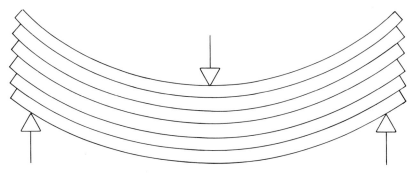

Fig. 4.2. *Independent bending of layers in horizontally sliced beam showing presence of horizontal shearing action*

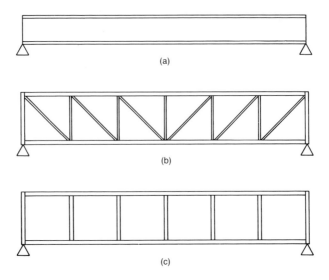

Fig. 4.3. *Examples of different beam forms: (a) I-beam, shear resistance provided by continuous plate or web; (b) lattice girder, shear resistance provided by series of braced panels; (c) Vierendeel girder, shear resistance provided by series of rigidly-jointed frameworks*

The element joining the flanges will support the compression flange against buckling in the plane of the beam, and, in order that it should resist the shearing tendencies, it must form a series of shear resistant panels of the type studied in chapter 2. Three possibilities are illustrated in Fig. 4.3. Counterbraced panels could also be used, and, again, these were popular in Victorian times and may be seen in many beam elements in the older railway station structures.

4.2   I-beams

An I-beam, as illustrated in Fig. 4.3(a), may fail in a number of ways, and an economical design requires that all secondary failure modes are eliminated in order that the full, in-plane, flexural strength of the beam may be developed. These secondary failure modes may be identified by testing a series of simple beams made from balsa wood and cartridge paper, and with the cross-section shown in Fig. 4.4

Fig. 4.4. *Cross-section of model I-beam*

A suitable test arrangement is shown in Fig. 4.5 in which the beam is simply-supported and loaded with a central point load applied to the top flange. The first failure mode develops at a relatively low load, and is illustrated in Fig. 4.6. The web of the beam is not sufficiently strong to transfer the concentrated loads to the support reactions and buckles. The beam will twist if the two ends buckle in opposite directions, which frequently occurs.

The simplest way to eliminate this mode of failure is to glue vertical stiffeners, in the form of short lengths of balsa wood, to each side of the web at each end above the support points. These stiffeners must be glued to the flanges and to the web plate and the load is transmitted to them by direct bearing and then shed into the web via the glue lines. Thus the load is distributed more or less uniformly into the web rather than as a concentrated application as was previously the case.

The centre of the web, beneath the central point load, is very slender and buckling of this region will next occur, again at a relatively low load,

Fig. 4.5. *Test arrangement for model I-beam*

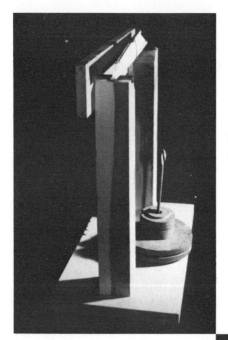

Fig. 4.6. *Twisting, buckling mode due to low stiffness of web*

Fig. 4.7. *Use of vertical stiffeners to support web at reaction points; central web buckling failure*

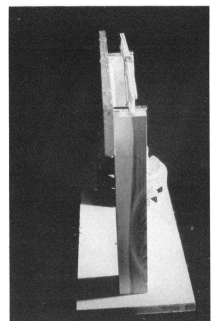

Fig. 4.8. *Lateral buckling of compression flange developing*

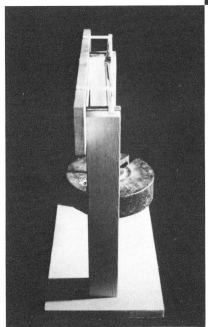

Fig. 4.9. *Ends of beam held in plane and lateral buckling developing at higher load than beam in Fig. 4.8*

Fig. 4.10. *Lateral buckling failure*

as shown in Fig. 4.7. Again this mode is best eliminated by glueing balsa wood stiffeners to each side of the web at the critical section. The stiffeners beneath the point load and above the two reactions are known as load-bearing stiffeners, for obvious reasons. An alternative to using load-bearing stiffeners would be to increase the thickness of the web plate and, in practice, this is frequently more economical for relatively lightly loaded beams.

Once the web has been stiffened against the effects of concentrated loads, the next mode of failure to develop is normally that due to lateral instability of the compression flange as illustrated in Fig. 4.8. The whole of the compression flange gradually moves sideways relative to the tension flange, thus causing the web also to deflect sideways, and the load to act eccentrically with respect to the line joining the two support reactions. Sudden collapse of the beam then follows.

For a rectangular section the weakest axis for buckling is that parallel to the longer side, as is evident if a compression load is applied to, say, a wooden ruler. In the case of the compression flange of the I-beam, however, the web plate prevents buckling in the plane of the beam, and it is not until a higher load is reached, sufficient to buckle the compression flange about the other axis, that collapse occurs. The buckling of the compression flange is not quite so simple as the buckling of columns studied in chapter 3, since the resistance the beam has to torsional deformations will also affect the buckling load, but this effect is relatively small with open sections such as the I-shape. Also, the compression load varies along the flange, following the pattern of the bending moment diagram, and is zero at each end. An example of lateral torsional buckling in a steel I-beam is illustrated in Fig. 1.16.

The simplest way to prevent lateral buckling of the compression flange is to provide it with lateral support at each end, and this may be achieved by glueing bracing strips between the ends of the compression flange and the back board of the loading rig. In this way the effective length of the compression flange is reduced to the span of the beam, thus allowing a greater load to be carried. Eventually, however, the centre of the compression flange will move sideways relative to its ends, as shown in Fig. 4.9, and collapse follows (Fig. 4.10).

Additional lateral support is therefore required at mid-span, and this should then be sufficient to prevent further problems with lateral buckling

Fig. 4.11. *Diagonal web buckling: load now carried by post-buckling tension field action; note how flanges are being pulled towards one another*

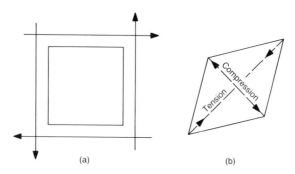

Fig. 4.12. *Effect of complementary shearing systems*

since the unsupported length has now been halved, and, as was seen from the studies in chapter 3, this produces a considerable increase in load-carrying capacity. However, the more lateral supports provided, the lower the chance of lateral buckling of the compression flange, and the ideal solution in practice is to use any available floor elements, such as concrete slabs, to provide continuous lateral support to the compression flange. The floor elements will be prevented from moving by the system of shear walls in the structure, as described in chapter 2.

If the model beam shown in Fig. 4.4 is manufactured with two paper webs, one on each side of the cross-section, rather than with the one central web, then the resulting beam is called a box beam. A box section has a very high resistance to torsional deformation and, consequently, lateral torsional buckling is unlikely to occur. This is one of the reasons for the popularity of box-beams in bridge construction.

As an alternative, if no lateral supports are available, lateral torsional buckling may be eliminated by increasing the size of the compression flange. One common solution in structural steelwork practice is to weld a channel section with the web horizontal to the compression flange of an I-section. Such a beam is called a compound beam.

Once lateral buckling of the compression flange has been eliminated, the next signs of distress to be observed are diagonal buckling folds in the web plate, as may be clearly seen in Fig. 4.11. The reason for these may be demonstrated by considering a small square element in the web of a beam at the centre of the depth. This will be acted on by vertical and horizontal shearing systems, as shown in Fig. 4.12(a), neither system being able to exist without the other in order that rotational equilibrium of the small element may be maintained.

These two complementary shearing systems cause the small element to deform in the manner shown in Fig. 4.12(b), and thus diagonal compressive and tensile forces develop as indicated. Therefore, four types of force are involved and, depending upon the material from which the beam is made, each could cause failure. Consideration of these four forces together with fundamental material properties helps to explain the forms of cross-section that are used for common beam systems, and why the efficient I-beam shape cannot always be used.

For example, timber is a material that is relatively weak in shear when this is acting along the grain. Consequently, the horizontal shearing forces can cause splitting along the grain in timber beams and such beams are normally made rectangular in section, with a relatively thick web, so that there is a large area of timber available at the neutral plane to help keep the horizontal shear stress low. If the relatively more efficient I-section is to be used in timber then the web must be fabricated in a form that is strong in shear. One good solution is to employ plywood, which is manufactured by peeling thin veneers of timber from a log and then glueing them into sheets, an odd number of veneers thick, with the grain directions alternately at right angles to one another. When shear is applied to a plywood sheet some of the veneers will be sheared parallel to the grain, and therefore contribute a small resistance, but the remainder will be sheared across the grain and these will provide a high resistance. By reconstituting timber in this manner a shear-resistant form is achieved, and

Fig. 4.13. *Use of intermediate stiffeners to hold flanges apart; post-buckling tension field action apparent*

by using it as the web of an I-beam, with the flanges formed by glueing solid timber on each side top and bottom, an efficient beam form results.

Alternatively, the web of an I-beam may be made up with diagonal boards so that the vertical and horizontal shears are not applied along the grain. Other forms of reconstituted timber sheet, such as oil-tempered hardboard, can also be used to form the web of a timber I-beam.

Turning to concrete as a material for beams, a fundamental weakness of this material is its low resistance to tension. Consequently, the diagonal tension force set up by the complementary shearing forces can cause problems and lead to a typical diagonal tension crack failure. Reinforced concrete beams are therefore normally made with a relatively thick web so that the resulting diagonal tension stress is kept low. However, by precompressing the concrete longitudinally, as in prestressed concrete beams, the precompression force combines with the diagonal tension force to produce very low or even no resultant tension, and hence I-beams with relatively thin webs may then be used.

In an I-beam construction with relatively very thin webs, such as in steel and aluminium plate girders, and in the model beams illustrated here, it is the diagonal compression force which is important and can cause buckling in the thin web element, as illustrated in Fig. 4.11.

It will be realized from the above discussion that, although the I-beam shape of cross-section is theoretically good for resistance to bending (owing to large areas of material being concentrated at the extreme surfaces) the effects of shear and the material properties can dictate the use of other

Fig. 4.14. *In-plane failure of beam at point of maximum bending moment*

shapes of cross-section. This is frequently acceptable in relatively cheap low-density materials, but would not be so with a material such as steel. In this case another solution must be found.

Referring to the model shown in Fig. 4.11, the web has buckled and can no longer resist the diagonal compression force. The load is still being carried, however, and consequently the beam must have developed a post-buckling support system. This may be likened to a lattice girder with tensile membrane forces acting along the buckle lines, and compression forces in the vertical stiffeners. This ability to develop a post-buckling membrane action, and hence to continue to carry load, shows a fundamental difference between the buckling of web plates in beams and the buckling of columns.

The tensile membrane forces exert an inward pull on the flanges of the beam, as may be seen in Fig. 4.11, and this eventually either leads to a flexural failure in the flanges or a tensile failure in the web. If more vertical stiffeners are provided to the web of the model, as shown in Fig. 4.13, then the flanges have more support and are better able to resist the pull of the tensile membrane forces; a higher load-carrying capacity will then be achieved. These additional stiffeners are called intermediate stiffeners. An alternative solution would be to use a corrugated web plate, in which case the fold lines can provide the necessary support.

Provided enough intermediate stiffeners are supplied, possibly accompanied by a small increase in web thickness if web failures were apparent, the model beam will now be able to carry a considerably higher load. Diagonal buckling of the panels will still be apparent, at a steeper angle than before, but the post-buckling membrane action will enable the full in-plane strength to be reached, as shown in Fig. 4.14.

The above sequence illustrates the approach that is used in the practical design of I-beams, whereby each secondary mode of failure is eliminated in order that the full strength of the section may be realized.

4.3   Lattice girders

An alternative beam form, which is more efficient than the I-beam when spans are larger and self-weight becomes important, is the lattice girder. It also has the advantage that the open nature of the web provides an easy passage for service ducts and pipes. In this type of beam the shear resistance is provided by the diagonal members acting in conjunction with the vertical and horizontal members.

Examples of simple lattice girder construction are illustrated in Figs 4.15

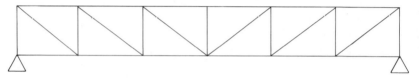

Fig. 4.15. *Lattice girder with tension diagonals*

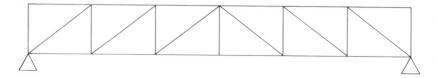

Fig. 4.16. *Lattice girder with compression diagonals*

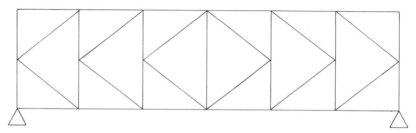

Fig. 4.17. *Lattice girder with K-bracing*

Fig. 4.18. *Glued timber I-beams with plywood webs*

Fig. 4.19. *Heavy welded steel plate girder in fabrication shop; load-bearing and intermediate stiffeners may be seen*

and 4.16. The arrangement shown in Fig. 4.15 is generally preferable, since the internal members which carry compressive loads (struts) are the shorter ones with consequently higher buckling capacities. The longer internal members are in tension (ties) and, since the load-carrying capacity of a tension member is independent of its length, this arrangement of short struts and long ties is particularly effective. The form illustrated in Fig. 4.16 reverses this principle and is, consequently, less efficient.

In any triangulated framework the principle of making the longer members ties and the short members struts should be followed when planning the arrangement in order to achieve maximum economy. Also, in order to keep the forces in the diagonal members at a reasonable level in relation to the applied loads the angles between the internal members should preferably be reasonably large, say between 40 and 60°.

If due to heavy loading, the depth of a lattice girder is made large (to provide a good resistance to bending and deflection) the vertical compression members will need intermediate support to help control buckling, and this may be achieved by the use of K-bracing, as shown in Fig. 4.17.

For relatively light loading the lattice girder may take the form of a series of equilateral triangles, with consequent advantages in terms of fabrication due to repetition of detail. Such a girder is called a Warren girder.

4.4   Practical examples

Figures 4.18–4.22 illustrate a number of practical beam elements.

Fig. 4.20 (below). *Nailed timber I-beam using diagonal boarding to provide shear resistance*
Fig. 4.21 (above right). *Timber I-beam with corrugated plywood webs*
Fig. 4.22 (below right). *Warren girder pipe bridge fabricated from hollow steel sections; overall length 146 m*

4.5 Exercises

Take a strip of cartridge paper 300 mm long and 25 mm deep, and glue 3 mm square matched balsa wood to each side of the top and bottom edges of the paper. This forms an I-beam with balsa wood flanges and a paper web. Support the beam on a span of 250 mm using an arrangement similar to that shown in Fig. 4.5.

Apply a central load to the top flange by hanging a loop of string over the centre of the beam and suspending the load from the loop. Provide packing to prevent the string from cutting into the balsa wood.

Gradually increase the load on the beam until it fails.

Note the mode of failure and the load at which it occurs and then devise a means of eliminating this failure mode. Build another model incorporating this modification and repeat the test. Again note the manner of failure and the failure load, and devise a means of eliminating this mode in the next model.

Continue in this way until the full, in-plane, flexural strength of the beam has been developed, all other secondary modes of failure having been eliminated.

Tabulate all failure modes and the loads at which they occur.

The experiments can be extended to study the behaviour of box beams. In this case glue two pieces of 3 mm square balsa wood together to form a piece 6 mm × 3 mm × 300 mm long. Make two such pieces and use them to form the top and bottom flanges of a box beam, the two webs being formed from strips of cartridge paper 300 mm long and 25 mm deep as before.

Repeat the test procedure as for the I-beam exercise.

Compare the results obtained from the two series of tests.

5 Arches and roof trusses

5.1 Introduction

If a length of string is tied between two anchor points and a weight is supported on the string, as shown in Fig. 5.1, then the string will deform until an equilibrium position is reached. Since string cannot resist compression, bending, shear or torsion, but only tension, then the resulting structure can be referred to as a tension structure. For the given position of the weight there is a whole family of tension structures, the actual equilibrium shape being dictated by the length of string used. It may be noted that the tension structure has straight members.

If a second weight is now added, in a different position, then the shape of the tension structure changes until a new equilibrium position is reached.

Fig. 5.1. *Simple tension structure carrying single point load*

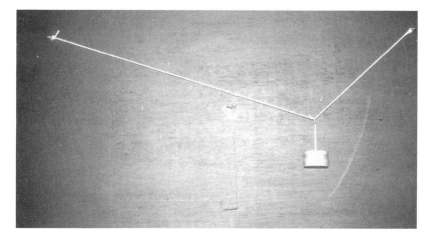

Fig. 5.2. *Simple tension structure carrying two point loads*

Once again there is a family of structures and the members are straight (see Fig. 5.2).

A uniformly distributed load may be modelled by using a chain, and if this is suspended between two anchor points, as shown in Fig. 5.3, then the corresponding tension structure will be curved; this curve is called a catenary since the load is distributed uniformly along the curve. If the load were distributed uniformly along the horizontal span then the corresponding

Fig. 5.3. *Tension structure for uniformly distributed load*

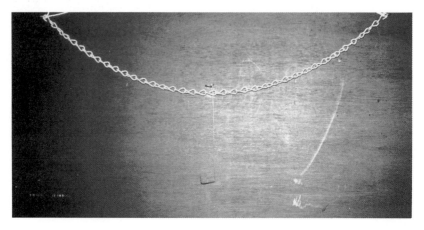

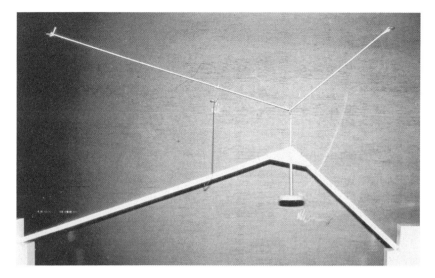

Fig. 5.4. *Inversion principle: string tension structure; balsa wood compression structure*

equilibrium curve would be a parabola.

Returning to the structure shown in Fig. 5.1, a balsa wood model may be constructed to the same shape as the string. If this is inverted, supported on two blocks and loaded with the same load as before, then the resulting structure will be a pure compression structure or arch. This is illustrated in Fig. 5.4. Since the balsa wood structure is in a state of pure compression, it would be possible to cut the members up into short lengths without damaging the integrity of the structure.

This inversion technique was used by some of the designers of the classical cathedrals. String models were built and loaded with weights which represented the loads that a roof element had to carry. By inverting the resulting form the shape of the required arch was obtained, which could then be built from individual stone blocks.

5.2 Arches

Assume that an arch structure is to be built to carry a central point load over a given span. The string model shown in Fig. 5.5 indicates the general shape that is required, and by the inversion technique the corresponding

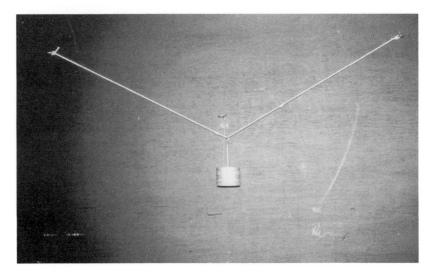

Fig. 5.5. *Simple tension structure for central point load*

Fig. 5.6. *Simple compression structure, or arch, for central point load*

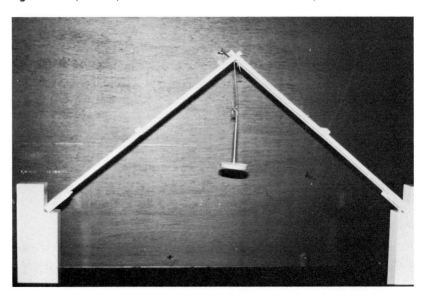

Fig. 5.7. *Instability of supporting structure*

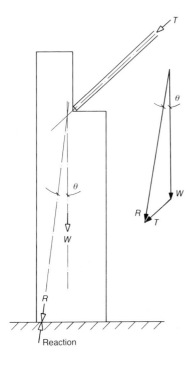

Fig. 5.8. *Forces acting on supporting structure*

arch is shown in Fig. 5.6. Lateral support is provided at the apex to keep the arch in plane.

If the central load were to be increased, however, overturning of the supports would eventually occur, as shown in Fig. 5.7. In order to rectify this situation it is necessary to examine the reasons for this form of instability.

The forces acting on the supporting wall element comprise a compression thrust T acting down the leg of the balsa wood arch and a vertical weight force W due to the self-weight of the wall. These two forces combine to produce a resultant force R, as shown in Fig. 5.8. If the resultant force falls within the base width of the wall then equilibrium will be maintained. However, if T is increased, due to an additional load being placed at the apex, R and θ will both increase, and if the resultant now falls outside the base then the wall will overturn (see Fig. 5.7).

One way of overcoming this problem would be to increase the width of the base of the wall so that the resultant would need to be deflected further before it fell outside the base. This solution could be expensive in practice and also could lead to sliding failures of the wall if the horizontal component of R exceeded the frictional resistance.

An alternative solution is illustrated in Fig. 5.9. In this case the value of W has been increased by applying weights to the top of the walls. This

Fig. 5.9. *Stability restored by increasing weight of supporting structure*

Fig. 5.10. *Stability restored by increasing pitch of arch*

Fig. 5.11. *Stability restored by propping supporting structure*

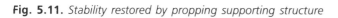

Fig. 5.12. *Tied arch solution; no overturning of supporting structure now possible*

has the effect of reducing the value of θ, and therefore it increases the likelihood that R will remain within the base dimensions of the wall. This is the principle of the buttressed wall in which the heavy weights of the buttresses stabilize the walls against the overturning thrust from the pitched roof members.

Another way in which the value of θ could be reduced, and hence stability restored, would be to increase the steepness of the arch. This approach is illustrated in Fig. 5.10 and is commonly adopted in church roof structures. Increasing the pitch of a roof produces an increased roof area, however, and hence additional cladding costs.

Yet another solution is shown in Fig. 5.11 in which the wall is propped against overturning. The example shown provides propping action from a weighted buttress, and this is the principle of the flying buttresses used in Gothic churches. The internal wall may then be reduced in dimension and this produces a more graceful structure. Another form of this solution would be to continue the line of the arch through the wall and into the ground, in order to achieve the necessary propping action.

Probably the most economical solution of all is shown in Fig. 5.12. In this case the bottoms of the arch members, or abutments, are tied together with a string. The string will resist the horizontal component of the arch thrust T and only the vertical component will be resisted by the wall. The wall is now only carrying vertical forces and there is no possibility of the resultant falling outside the base, so stability is guaranteed. As the load is increased the compresssion force in the arch and the tension force in the string will both increase until the arch member buckles like a column, or the string breaks. This system is called a tied arch and is commonly adopted in cases where the ties do not obstruct the proposed use of the building.

5.3   Roof structures

So far the various types of arch that have been represented have only been loaded with a central point load. In a roof structure this is unrealistic since loads will be distributed all over the roof due to the self-weight of construction, snow loading and wind loading. Consequently, more than one point load will be applied to the principal structural elements. A typical arrangement is shown in Fig. 5.13 in which the loads on the roof cladding are carried onto purlin beams which then transfer the loads as point loads to the principal elements.

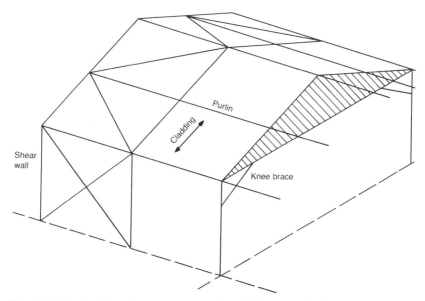

Fig. 5.13. *Typical framing arrangement for single-storey shed*

Fig. 5.14. *Typical loading pattern on principal roof element*

When considering the two-dimensional principal elements it is important to ensure that they are stabilized in the third dimension, otherwise they could easily fail out of plane. This is frequently achieved in practice by the use of plate or truss elements to carry the forces in the third dimension to a system of shear walls. For example, in a roof structure the purlins will maintain the principal roof elements in plane, provided they are themselves prevented from moving by a triangulation system in the plane of the roof, or by the roof cladding acting as a plate, either being supported in turn by a shear wall system (Fig. 5.13). The building is thus stabilized longitudinally, and lateral stability may conveniently be provided by the use of knee braces to form rigidly jointed frames.

A typical arrangement of loads on a principal roof element will, therefore, be as shown in Fig. 5.14. However, this shape of roof, which is very convenient for drainage, is no longer following the inversion technique, and consequently will not be in a state of pure compression. Fig. 5.15 shows the discrepancy between the actual shape and the inverted tension structure shape, or line of thrust as it is sometimes called.

The main arch members are now subjected to a combination of

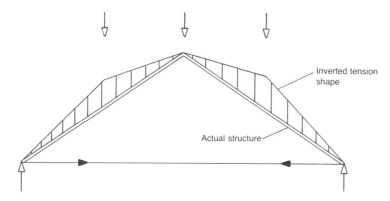

Fig. 5.15. *Deviation between triangular tied arch and inverted tension structure shape for loading pattern shown in Fig. 5.14*

Fig. 5.16. *Use of internal strut—tie combinations to remove bending effect from intermediate loads, thus forming Fink roof truss*

Fig. 5.17. *Fink trussed rafter commonly used at close centres to form domestic roof structures*

compression and bending, the amount of bending being proportional to the deviation between the actual structural shape and the inverted tension structure. Structures can be designed to resist these combined effects but, in general terms, it is more efficient for a member to carry only tension or compression loads.

The bending effect in the model shown in Fig. 5.14 could be eliminated if some propping action were provided to the two loads applied at the quarter span positions. Propping to the ground would obviously work but it would not be very practical, and a prop applied beneath each quarter span load and the string would not work since the string would be pushed downwards. However, if the loads were propped onto the string and the joints on the string tied up to a stiff point, then the desired result could be achieved. The stiffest point available is at the apex of the arch, and this arrangement of propping and tying is illustrated in Fig. 5.16. The resulting structure is called a Fink roof truss. The balsa wood members will all carry compressive loads and the string members tensile loads. A timber Fink truss is illustrated in Fig. 5.17.

If the span of a roof truss were to be increased, more point loads would be applied to the truss by the purlin system, and thus more internal propping and tying would be necessary, so forming complex truss forms such as that shown in Fig. 5.18.

Fig. 5.18 (right). *Tubular steel trusses for factory building 18·5 m span*

Fig. 5.19. *Tubular steel trusses for greenhouse 21·5 m span*

5.4   Practical examples

Figures 5.19 and 5.20 illustrate other arch and truss structures.

5.5   Exercises

Construct a series of tied arch models to span 150 mm, each with a different pitch angle ranging from 15 to 75°. Make the sides of the arch from pairs of 1·5 mm square balsa wood sticks spaced 3 mm apart and glue 10 mm × 6 mm card spacers across them at each end and at the centre, to hold them in position.

Fig. 5.20. *Glued laminated timber tied arch roof*

Shape the top ends so that they bear against one another and glue small card gussets on each side to form the top joint.

Use string to form the tie member passing each end through a hole in the lower card spacer. Knot the string at each end and glue it to the card spacer.

Support each arch in turn in a manner similar to that shown in Fig. 5.6. Keep the arches in plane by providing lateral support at the top.

Hang a load from the top of each arch and gradually increase the load until failure occurs.

As the pitch angle is increased, the members of the arch are better aligned to the direction of the applied load and are therefore able to carry a higher applied load.

On the other hand, the length of the side members increases as the pitch angle increases and this will reduce the buckling capacity, as was observed from the work described in chapter 3.

The two effects interact and produce an optimum angle for the arch. Plot the ultimate load for each arch against the pitch angle and determine the optimum angle.

6 Grids

6.1 Introduction

If a floor system consists of a series of parallel beams all spanning in the same direction, as shown in Fig. 6.1, then each beam acts more or less independently of the others so that when a concentrated load is applied to one of them the individual beam needs to be strong enough to transmit the whole of the load to the supports. Also, the loads on the floor are almost entirely carried by the two supporting elements perpendicular to the beams, and any potential supporting elements parallel to the beams are wasted.

A much lighter form of construction would result if concentrated load applications could be shared between several members, so that no one member had to do all of the work, and if all the supporting elements could be utilized. A grid structure such as that shown in Fig. 6.2 is one way in which these aims can be achieved. In order that all beams may assist in carrying any applied loads through to the supports, it is essential that they are interconnected at each intersection point.

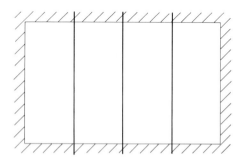

Fig. 6.1. *One-way spanning beam system*

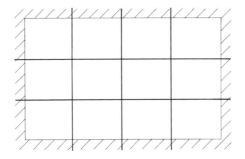

Fig. 6.2. *Two-way spanning beam system or grid*

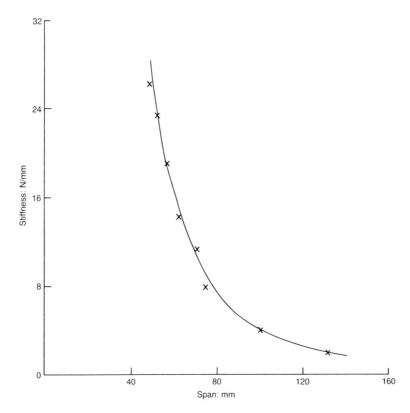

Fig. 6.3. *Effect of span upon stiffness*

A grid structure, therefore, consists of two or more intersecting parallel beam systems, each system being interconnected. The direction of the beam systems need not, however, be parallel to the support elements.

6.2   Stiffness

Stiffness may be defined as the load to cause a unit displacement, i.e. it provides a measure of the resistance offered to displacement.

Consider a balsa wood beam simply supported at each end and loaded with a central point load. By measuring the central deflection of the beam for a series of point loads, all within the capacity of the beam, a graph

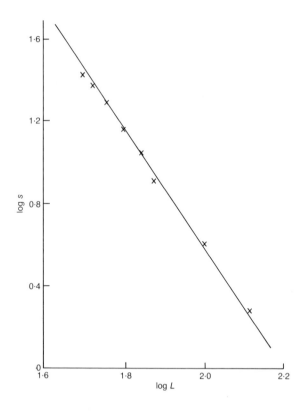

Fig. 6.4. *Logarithmic plot of results in Fig. 6.3*

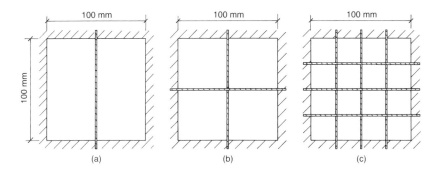

Fig. 6.5. *Three support systems for central point load: (a) single beam; (b) pair of equal length beams; (c) square grid*

may be plotted of load against displacement. The slope of this graph will represent the stiffness of the beam for central point loading.

By repeating this process for different spans of beam made from matching balsa wood (i.e. similar weight per unit length), a graph may be plotted of stiffness against span. A typical set of results is shown in Fig. 6.3.

If it is assumed that an exponential relationship exists between stiffness and length for a given beam, i.e. that $S \propto L^n$, then, as in chapter 3, a log−log plot, in this case log S against log L, will give the power index n. This is shown in Fig. 6.4, from which it can be seen that $n = -3$.

It may be concluded, therefore, that the stiffness for central point loading of a given size of simply supported beam, made from a given material, is inversely proportional to the cube of the span. Therefore, for example, halving the span of a beam increases its stiffness eight times.

6.3   Square grids

Figure 6.5 shows three possible support systems for a point load applied centrally to an area that is 100 mm square. In (a) the load is carried by a single beam; in (b) by a pair of beams forming a simple grid; and in (c) by a more complex grid system.

By making each of these structures from 3 mm square matching sticks of balsa wood, experiments may be carried out, similar to those in section 6.2, to measure the stiffness of each system under central point loading. Fig. 6.6 shows the more complex grid system under test.

Fig. 6.6. *Square grid under test*

A typical experiment gave the stiffnesses shown in Table 1. In model (*b*) the centre of each beam must be deflected by the same amount; let this be Δ. If the stiffness of one beam is S_1 and that of the other beam is S_2, then the force needed to deflect the first beam is ΔS_1 and that to deflect the second is ΔS_2. Therefore the total load needed to deflect the pair of beams is

$$\Delta S_1 + \Delta S_2$$

$$= \Delta(S_1 + S_2) = P, \text{ say}$$

The stiffness of the pair of beams for central point loading is by definition

$$P/\Delta = S, \text{ say}$$

$$\therefore S = S_1 + S_2$$

Thus the stiffness of the combined system is equal to the sum of the stiffnesses of the two beams.

Also, the load on the beam with stiffness S_1 will be

Table 1

Grid	Stiffness: N/mm
Fig. 6.5(a)	1·5
Fig. 6.5(b)	2·7
Fig. 6.5(c)	5·5

$$\Delta S_1 = \left[\frac{P}{S_1 + S_2}\right]S_1 = \left[\frac{S_1}{S_1 + S_2}\right]P$$

and that on the other beam will be

$$\left[\frac{S_2}{S_1 + S_2}\right]P$$

In model (b) each beam has the same stiffness, equal to that of the beam in model (a); therefore, model (b) should be twice as stiff as model (a). The experimental ratio is $2\cdot7/1\cdot5 = 1\cdot8$.

In model (c) the additional beams by virtue of their stiffnesses offer partial support to the central pair of beams. The spans of the central beams are therefore effectively reduced, and hence the stiffness of the system is increased. The load to cause failure would also be increased relative to model (b), since its effect is shared between so many members.

A one-way spanning beam system equivalent to model (c) would comprise three parallel beams. Since each beam would act independently, the stiffness of this one-way system, under a central point load, would be the same as that obtained for model (a). The one-way system would, however, use only half the amount of material, and hence for the grid construction of model (c) to show an advantage it should be at least twice as stiff.

From the experimental results the stiffness ratio is

$$\frac{(c)}{(a)} = \frac{5\cdot5}{1\cdot5} = 3\cdot7$$

6.4   Diagrids

If the beams forming the grid are arranged on a diagonal pattern, as shown in Fig. 6.7, approximately the same amount of balsa wood is used as for grid (c) in section 6.3 but an increase in stiffness will result, provided the ends of the longer diagonals are prevented from lifting. This increase will be typically about 20% for the arrangement shown; such a grid is called a diagrid.

The increase in overall stiffness is due to the relatively high stiffnesses of the shorter diagonal members which enable them to provide considerable support to the longer diagonal members. This, together with the need to hold down the ends of the longer diagonals, causes a reversal of curvature in the longer diagonals, thus reducing their central deflection and hence increasing their stiffness. In a rectangular grid the inter-supporting effect is less marked, and consequently reversal of curvature does not occur in any of the members.

In both grids some torsional deformation is apparent, and therefore the bending of each element is slightly modified by the torsional stiffness of the other elements.

The above differences in behaviour may be explained in another way by

Fig. 6.7. *Diagonal arrangement of beams or diagrid*

considering the pattern of isostatics, or principal stress directions, in a plate element. Fig. 6.8 shows the isostatic pattern for a plate simply supported on each side and carrying a uniformly distributed load. In the diagrid arrangement the beams more nearly follow the lines of stress, and consequently they are used more efficiently. In fact a grid structure may be looked upon as a degenerate plate form, and by making the beams follow the isostatic lines the maximum efficiency would be achieved. Fabrication costs would, however, be high and the diagrid provides a reasonable compromise.

If beams are placed in three main directions, the grid becomes a three-way grid with further advantages in terms of stiffness and strength.

If the interconnecting beams are replaced by lattice girders, a double layer grid structure results; this is a structural form capable of spanning large distances without intermediate supports. Also, space grids may be constructed by interconnecting triangulated space forms such as tetrahedra, and very large spans may be economically covered in this way.

Fig. 6.8. *Isostatic pattern, showing directions of principal stresses, for simply supported square plate carrying uniformly distributed load*

6.5   Practical examples

Figures 6.9–6.11 illustrate a number of practical grid structures.

Fig. 6.9. *Precast post-tensioned concrete diagrid roof*

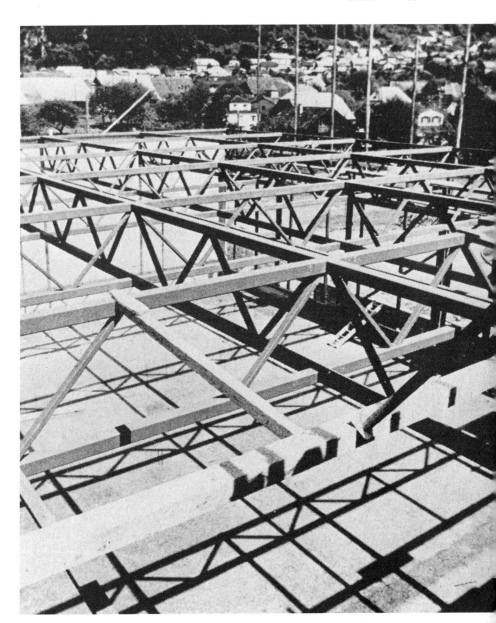

Fig. 6.10. *Two-way double-layer grid, 25 m × 24 m, forming roof for school sports hall; each intersecting lattice girder is fabricated from rectangular hollow steel sections*

6.6   Exercises

Take a piece of 3 mm square balsa wood and support it as a beam on a span of 125 mm. Load the beam centrally by suspending weights from a loop of string and determine the load to cause a unit displacement, i.e. determine the stiffness for mid-span loading. Any convenient unit displacement may be taken, but the loads must of course be within the capacity of the beam.

Repeat the test for other spans down to 50 mm, using matching balsa wood, and plot a graph of stiffness against $1/(span)^3$.

What shape of graph should be obtained?

Make a list of the small errors in the experiment which could contribute to a slight scatter in the points.

Using the same quality of 3 mm square balsa wood as in the above exercise, take two pieces, one 125 mm long and the other 75 mm long. Glue their mid-points together, placing the 75 mm length over the 125 mm length.

Simply support the four ends, and use a loop of string to suspend a small weight from the centre of the system. Determine the stiffness for mid-span loading.

Compare the result with the sum of the stiffnesses, obtained from the exercises above, for the two members acting independently.

Fig. 6.11. *Space grid fabricated from rectangular hollow steel sections, 183 m × 60 m overall and supported by 16 main columns*

7 Folded plate structures

7.1 Introduction

Referring back to the roof system illustrated in Fig. 5.13, there is an alternative structural form which could be used if the roof were to be treated as a whole rather than as a series of plane elements.

Consider the roof structure shown in Fig. 7.1. The load at point B on the ridge may be resolved into two components acting down the planes of the roof. These component forces must be resisted by some means, or the roof will spread at eaves level. One of the solutions described in chapter 5 provided for a tie member joining A and C, and for vertical reactions at A and C. If, however, the inclined sides of the roof are made into beam

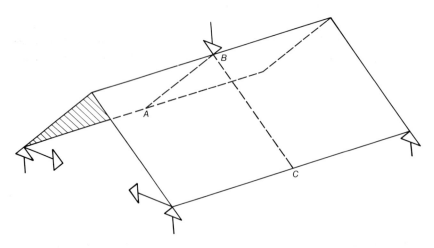

Fig. 7.1. *Basic folded plate roof form*

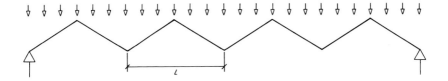

Fig. 7.2. *Multiple folded plate roof form*

or girder elements, then the component forces may be carried by beam action to the ends of the structure where the necessary in-plane reactions may be provided by combined vertical support and horizontal tie systems. Any tendency for lateral instability to develop in the compression flange of a beam element is resisted by the supporting action of the adjacent beam element.

The above principles may be extended to a multiple roof form such as that shown in Fig. 7.2. In this case the more general condition is illustrated where the loads are applied between the eaves and ridge levels; for example, those due to snow and the self-weight of the structure. These loads have, therefore, to be conveyed to the ridge and eaves lines by the transverse bending strength of the beam elements, before the longitudinal beam action described above can transmit them to the supports. At each intersection, or fold line, support is provided by the longitudinal action of the two inclined beam elements, so the transverse span of the elements is only $L/2$ and can be made smaller by the use of more fold lines, as shown in Fig. 7.3. Therefore the transverse bending strength of the beam elements does not need to be large. Since only one inclined beam element is available

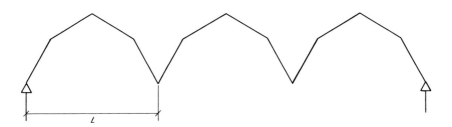

Fig. 7.3. *Use of additional fold lines to reduce transverse span of beam elements*

Fig. 7.4. *Low stiffness of thin slab element*

at the extreme edges, continuous support must be provided there in the form of vertical beam or wall elements.

The longitudinal beam elements can take any of the beam and girder forms illustrated in chapter 4, provided they have the necessary transverse strength to span between fold lines.

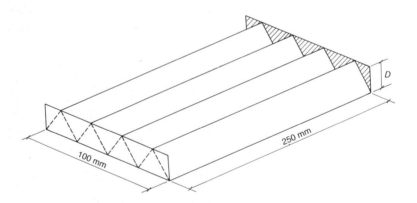

Fig. 7.5. *Folded plate model*

The structural form described above is commonly called a folded plate because of its appearance, although the expression 'stressed skin construction' is also used, since it is descriptive of the way in which it functions.

7.2   Simple model forms

If a piece of thin card is supported on two sides and a load applied to it, as shown in Fig. 7.4, then considerable deflection will occur. This simple slab element does not have a high inherent stiffness. This could be countered by using a thicker piece of card or, more efficiently, by using a stiffer structural form, such as a folded plate structure.

A simple folded plate structure is illustrated in Fig. 7.5 and this may be manufactured from thin card. If each fold line is lightly scored with a ballpoint pen before folding, a clean edge will result. Thin card diaphragms may be glued to each end to provide the resistance to spread under load. The effectiveness of this simple structure is illustrated in Fig. 7.6.

If the depth of the roof form D is made too low, then the geometry of the folded plate structure will change at mid-span owing to horizontal

Fig. 7.6. *Folded plate model under test demonstrating high degree of stiffness*

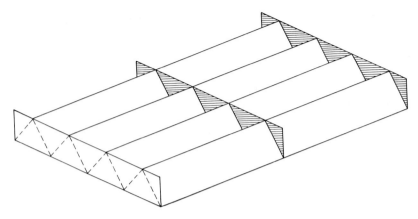

Fig. 7.7. *Folded plate model with central diaphragm*

spreading, and it is possible for the folds to flatten, so producing a 'push through' type of failure.

If the depth of the corrugations were to be increased, then the inclined beam elements would be better aligned with respect to the direction of the applied load, and consequently more load could be carried for a given in-plane strength. At the same time, however, the in-plane depth of the beam elements would increase and, although this would raise the bending strength, the danger of web buckling failure would be made more severe. This interaction of better alignment of the elements and increasing buckling tendency leads to the concept of an optimum depth of corrugations for a given material and span.

If a diaphragm were provided at mid-span, by glueing triangular pieces of thin card to the upper surfaces as shown in Fig. 7.7, or to the under surfaces, then changes in geometry at mid-span could be controlled and the diaphragm would also act as a stiffener to the beam elements at the points of concentrated load application, thus controlling the web buckling tendency. Higher strengths could be achieved with these modified models, and upon loading to failure other modes of failure would appear, similar to those found in the I-beam study.

Space-enclosing folded plate structures may be manufactured, and Fig. 7.8 shows how to fold a sheet of thin card to obtain one of these forms. The card is lightly scored with a ball-point pen on one side along the ridge fold lines and then on the other side along the valley fold lines. It may then be folded in 'concertina' fashion and upon reopening the structural form will become apparent. By glueing it to a base board a relatively very stiff structure will be obtained, considering the thin material from which it is made, as shown in Fig. 7.9. This high stiffness comes from the shape of the structure rather than from the stiffness of the material, and for this reason folded plate forms are very suitable for low modulus materials such as plastics and timber. An ideal combination would use sandwich construction (see chapter 8) to give the elements transverse bending strength and stiffness, and folded plate construction to give the structure overall strength and stiffness.

Further examples of folded plate structures made from paper are shown in Fig. 7.10.

7.3   Practical examples

Figures 7.11−7.14 illustrate a number of practical folded plate structures.

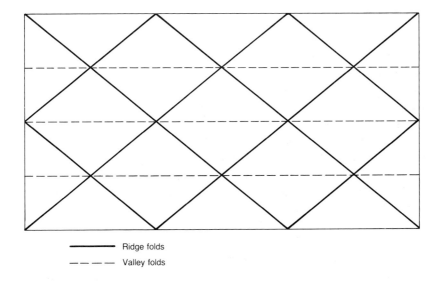

———— Ridge folds

— — — — Valley folds

Fig. 7.8. *Arrangement of fold lines to produce space-enclosing structure*

Fig. 7.9. *Type of structure obtained using procedures illustrated in Fig. 7.8*

Fig. 7.10. *Other forms of folded plate models*

Fig. 7.11. Folded plate roof in plywood and timber

Fig. 7.12. Concrete folded plate roof for grandstand

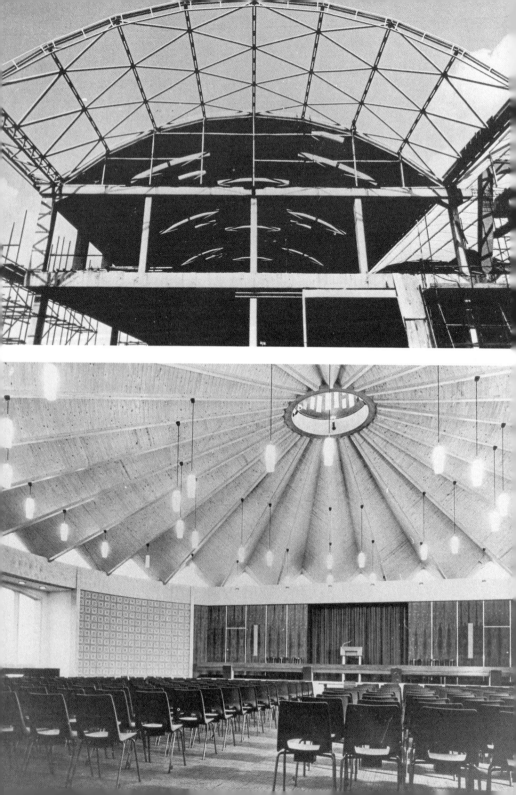

7.4   Exercises

From matching-quality sheets of cartridge paper, construct a series of simple folded plate models to the dimensions shown in Fig. 7.5. The depth D of the corrugations is to vary from model to model between the limits of 10 and 60 mm. The number of folds is, however, to be kept constant. In order to maintain the shape of each model, and to provide resistance to spread under load, glue thin card diaphragms to each end.

Simply support each model in turn on a span of 250 mm and apply loads to the centre of the two middle ridge lines. This may be achieved by placing a piece of wood between the two loading points and applying weights to the middle of the wooden cross-member.

Continue to increase the load on each model until failure occurs. Note the mode of failure.

Plot a graph of ultimate load against depth D of corrugations and determine the optimum depth for this model.

Fig. 7.13 (above left). Folded plate, barrel vault roof for transit shed fabricated from rectangular hollow steel sections; seven bays, each 23 m wide and 53 m long, including 9 m cantilever
Fig. 7.14 (left). Circular plan, radial folded plate church roof in timber

8 Composite behaviour

8.1 Introduction

The expression 'composite behaviour' is normally used in practice to mean either the working together of two materials to form one element, as in the case of reinforced concrete, or the working together of two elements to form a new combined element. In general, composite action leads to greater economy in the use of materials.

A natural extension of this idea would be to obtain composite action between all the elements that make up a structural form, and to design the structure as one complete body. Such composite action does undoubtedly occur, to some degree, in many structures but a lack of data prevents all but a very few from being treated in this way, hence the reason for dividing the structures into elements for design purposes.

8.2 Beam and slab systems

A model of a beam and slab system is illustrated in Fig. 8.1. It comprises three balsa wood beams, manufactured by glueing together pairs of 3 mm

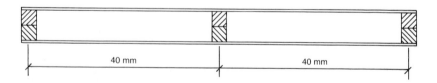

40 mm   40 mm

Fig. 8.1. *Cross-section of beam and slab model*

Fig. 8.2. *Test arrangement for models*

square matched balsa wood sticks, and two slabs of thin card; the models were made 175 mm long.

If three of these models are tested as shown in Fig. 8.2 with each having different degrees of connection between the beam and slab elements, then major differences in performance are found. These differences are illustrated by the typical set of results shown in Fig. 8.3. In this experiment three degrees of connection were investigated

(a)   two slab elements not glued to the beams

(b)   top slab element glued to the beams, the bottom element unglued

(c)   both top and bottom slab elements glued to the beams (sandwich panel).

It may be observed from Fig. 8.3 that at low loads each model exhibited a linear relationship betwen load and deflection, but that at higher loads the relationship became nonlinear. In practice, the loads that are applied to structures must be well below the ultimate load for safety, and consequently a structure is normally working within the linear portion of its characteristic.

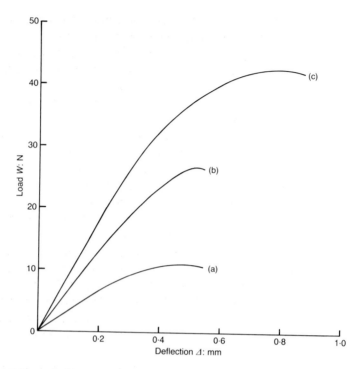

Fig. 8.3. *Typical load–deflection characteristics*

For these models the stiffness for central point loading (see section 6.2) will be represented by the slope of the characteristic. If from Fig. 8.3 the stiffness of model (a) is taken as unity, then the stiffness of model (b) is 1·8 and that of model (c) is 2·5. These increases in stiffness have, therefore, been produced by glueing the elements together, thus enabling them to work compositely. This composite action is only possible if the glue lines are capable of resisting the horizontal shearing forces that are developed.

Model (a) has no glue, and therefore composite action does not occur, the slab elements slipping relative to the beam elements in a manner similar to that illustrated in Fig. 4.2.

In simple terms, the effect of glueing the slab elements to the beam elements has been to enable a portion of the slab to act in conjunction with each beam, so that the rectangular beams of model (a) have become T-shaped beams in model (b), and I-shaped beams in model (c). Thus the increases in stiffness have been achieved by providing more material at the extreme surfaces where the maximum bending strains occur.

These stiff shapes are particularly important if materials with a low inherent stiffness (low Young's modulus) are to be used successfully in practical structures. Their use counteracts the low stiffness of the material to produce an element which will not deflect excessively. Consequently, the sandwich panel is frequently used in timber construction. Nailed sandwich panels can also be constructed in timber as an alternative to glued construction. In this case what is called 'incomplete composite action' will occur. This is because a nailed joint must slip before it can resist load, the amount of slip increasing as the load increases. Consequently, since the nails must resist the horizontal shearing forces between the slab and beam elements, some slip will occur at the interfaces and the resulting form will be less stiff than the corresponding glued form.

Another parameter that is of considerable importance to the designer is the ultimate strength of a structure, and Fig. 8.3 shows that model (b) has about 2·4 times the strength of model (a), and model (c) about 4·0 times the strength of model (a). Thus the composite action produces a considerable gain in strength as well as in stiffness. This is again because of the change in the effective shapes of the beams, the T- and I-shapes being strong shapes as well as stiff shapes.

Therefore, with composite action a given structural arrangement may carry more load, or conversely a smaller amount of material may be used to carry a given load.

The failure modes in the models are normally flexural, but occasionally horizontal splitting failures are observed in the balsa wood beams. These are due to the horizontal shearing forces which are, of course, present in the balsa wood as well as in the glue lines, and timber is particularly weak in shear parallel to the grain.

Some buckling of the thin card slab elements is also frequently observed, caused by the in-plane shear effects combining with longitudinal bending effects to produce compressive forces. Because the slab elements are thin they will normally buckle rather than crush.

The idea of a sandwich panel can be taken further. The two extreme surfaces are ideally arranged to carry bending effects since they are positioned at the levels of maximum bending strain. The intermediate element need only be strong enough to carry the shearing effects, and thus to ensure the composite behaviour of the two extreme surfaces, and to support the compression surface against buckling. (This is identical to the

Fig. 8.4. *Composite construction used for motorway bridge; stud connectors welded to top flanges of steel beams may be seen, and shuttering for concrete deck is being prepared*

function of the web plate in the I-beam element studied in chapter 4.) By making this intermediate element continuous and lightweight, a panel with a very high bending strength to weight ratio may result. This structural form is frequently used in aircraft structures, the surface elements commonly being aluminium alloy plates and the intermediate elements a foil honeycomb.

In Fig. 8.4 steel I-beams and box girders may be seen with stud shear connectors welded to the top flanges. The reinforced concrete slab is cast on top of the steel beams and the function of the stud connectors is to resist the horizontal shearing tendency at the interface, thus enabling some of the concrete slab to work as additional flange material. An increase in strength and stiffness will then be achieved. The heads on the stud connectors are necessary so that the reinforced concrete slab is tied down to the steel beam against any vertical buckling that might result from the compression forces in the slab.

In reinforced concrete construction, beam and slab elements are frequently cast monolithically, thus developing a T-beam shape. Part of the slab acts as a flange, thus giving a greatly increased compression area compared with that of a rectangular cross-section; this, in conjunction with more steel reinforcement in the tension area, enables a higher moment to be carried and greater stiffness to be obtained for a given depth of construction.

In timber roof or floor systems, composite action may be obtained by glueing plywood slab elements to the longitudinal timber framing members, either to the top surfaces, in which case the element is normally referred to as a stiffened panel, or to both top and bottom surfaces, thus forming what is called a stressed skin or sandwich panel. This method of construction is illustrated in Fig. 8.5.

Fig. 8.5. *Fabrication of composite plywood–timber flooring panel*

8.3   Exercises

Take a piece of thin card 50 mm wide and 200 mm long and support it over a span of 200 mm. Note its flexibility, i.e. lack of stiffness.

Using a suitable adhesive glue this card, and another similar one, to the top and bottom surfaces of a slab of polystyrene with the same plan dimensions; use a polystyrene thickness of about 10 mm. Support this sandwich panel on a span of 175 mm and measure its stiffness under a central load.

Repeat the experiment with a double thickness of polystyrene.

Compare the resulting stiffnesses.

9 Miniature bridge design exercise

9.1 Introduction

A useful way to integrate the knowledge gained by the studies outlined in the preceding chapters, and in particular to take account of the three-dimensional nature of actual structures, is to design and build a miniature bridge.

In order that the exercise should reflect the actual design process, the original design should be carried out on paper and no major changes should be made during the construction process. A limited amount of material and small component testing should provide the necessary design data. Strength and serviceability requirements should be specified.

Again, the learning process would be greatly enhanced if small-group working is allowed, although the exercise could be carried out by an individual.

9.2 Example exercise — miniature lifting bridge

Design, construct and load-test the superstructure of a miniature bridge to carry both vertical and horizontal loads of specified magnitude.

The bridge is to form the main structure of a lifting bridge, which in the whole structure would be combined with approach viaducts to form a crossing for a river estuary.

The deck of the bridge is to be constructed from a number of lengths supported on the test-loading module, half of which is shown in Fig. 9.1.

A length of deck on each end of the bridge will be simply supported by a dowel A at one end and a tower B at the other end. Each tower will stand freely on the module at datum level; the centre of each tower will be 300 mm, measured horizontally from the axis of the dowel.

The centre part of the bridge, between the two supporting towers (spaced 900 mm apart centre-to-centre) must be capable of being lifted. This may be achieved in either one of two ways

(a) by making the centre part in two halves, each pivoted at a support tower

(b) by making the centre part in one piece capable of being lifted vertically, guided in each of the support towers.

No fixings are to be made to the support module, in order that the whole bridge may be removed to enable the total weight to be measured. The dowel at A is a loose fit and can be removed for this purpose.

The deck must be 1500 mm long overall, have a maximum gradient (when

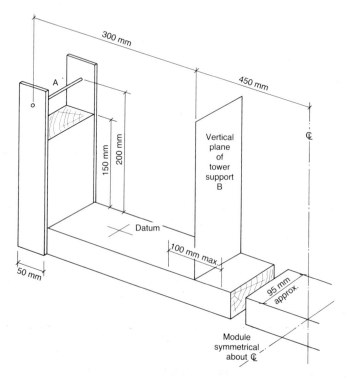

Fig. 9.1. *Support module for miniature bridge*

unloaded) of 10%, and a continuous surface. The deck should have a minimum clear width of 75 mm and a maximum overall width of 90 mm throughout. A minimum headroom over the clear deck width of 50 mm shall be provided throughout.

The width of the support towers at B, measured parallel to the span, should not exceed 100 mm.

When the bridge is in the lifted position, a clear space not less than 600 mm long (measured parallel to the deck) and 450 mm high (measured above datum level) must be provided beneath the central span of the bridge.

Under load condition 1 (see below), a clear space not less than 750 mm long (measured parallel to the deck) and 180 mm high (measured above datum level) must be provided beneath the central span of the bridge.

No mechanism for lifting the bridge is required — it may be raised by hand.

Dimensional tolerances are for spans ± 5 mm and for others ± 2 mm.

Materials for construction are limited to balsa wood, paper, cotton and balsa cement.

The bridge is required to satisfy separately each of the three loading conditions stated below.

1. *Vertical load.* A total distributed working load of 8·25 kgf applied over the full length of the deck. This can be achieved by a single layer of 12 mm dia. steel bars, 75 mm long, placed so as to touch one another and completely cover the deck.

2. *Wind load.* To simulate wind loading a horizontal point load of 0·3 kgf applied at the highest point of each support tower B. This may be achieved by attaching strings to the tops of the towers which pass over pulleys, each with a 0·3 kg mass attached. In practice the self-weight of the deck of a bridge is a large proportion of the total load supported. Therefore, to simulate practical conditions, apply the wind load with all of the vertical load from load condition 1 in place.

3. *Collapse load.* Design the bridge to give structural collapse at a vertical load of between 12·0 and 15·0 kgf uniformly distributed, with no wind load.

All stated vertical loads are in addition to self-weight. Test loading may be used to determine the strength and behaviour of elements where this cannot be calculated. However, the maximum dimension of any test element should not exceed 150 mm.

9.3 Comments

An exercise such as the one specified above models the structural design process quite well, although the resulting structures may not be wholly realistic.

The main virtue of the exercise is considered to be its educational value, and it is suggested that it be carried out by small groups of students (3 or 4 per group) working full-time for a week. The paper design and sketches should be finalized by the middle of the week to allow sufficient time for construction and testing.

If assessment is needed, this can be achieved both for the group and individual. Group assessment may be based upon

(a) an inspection of the completed structure and the design calculations and sketches
(b) the test loading of the structure
(c) the weight of the structure, to provide an approximate measure of cost.

Individual assessment can be achieved by requiring each student to write a critical appraisal of the exercise.

Index

Arches, 63
Battened columns, 41
Beams, 46, 54
Bending moment, 11
Box beams, 53
Buckling, 15, 19, 31, 38, 47
Buckling load, 35
Buttresses, 70
Catenary, 64
Closed sections, 40
Columns, 31
Composite behaviour, 100
Compound beam, 53
Compression, 7
Counterbraced panels, 23, 48
Couples, 8
Crushing strength, 15, 31
Design process, 1
Diagonal compression, 55
Diagonal tension, 55
Diagrids, 83
Effective length, 35
Effective length factor, 36
End fixity, 33
Equilibrium, 9, 68
Fixed end, 33
Folded plates, 90

Force, 5
Flying buttresses, 70
General stability, 18
Grids, 78
Horizontal shear, 47
I-beams, 48
Intermediate stiffeners, 57
Inversion technique, 65
K-bracing, 59
Knee braces, 72
Lateral torsional buckling, 18, 19, 53, 91
Lattice girders, 57
Load-bearing stiffeners, 52
Local buckling, 15, 38, 40
Moment of force, 7
Moment of resistance, 12
Multi-bay structures, 27
Multi-storey structures, 26
Pinned end, 33
Portal frame, 25
Post-buckling membrane action, 57
Principal stress, 13
Purlins, 72
Reaction, 5
Roof truss, 63

Sandwich panel, 101, 103
Second moment of area, 15, 35
Shear, 10
Shearing force, 11, 25
Shear resistance, 12
Shear walls, 19, 36, 54
Slender columns, 31
Square girds, 81
Stiffness, 80, 103
Stress, 12
Stressed skin construction, 93, 105

Stud connectors, 104
Temporary bracing, 27
Tension, 6
Tension structures, 63
Thin-walled columns, 36
Tied arch, 72
Torsion, 11, 84
Vertical shear, 47
Wall elements, 19
Warren girder, 59
Weight, 5
Young's modulus, 15, 35, 103